Interactive Graphics
for Data Analysis

Principles and Examples

Chapman & Hall/CRC
Computer Science and Data Analysis Series

The interface between the computer and statistical sciences is increasing, as each discipline seeks to harness the power and resources of the other. This series aims to foster the integration between the computer sciences and statistical, numerical, and probabilistic methods by publishing a broad range of reference works, textbooks, and handbooks.

SERIES EDITORS

David Blei, Princeton University
David Madigan, Rutgers University
Marina Meila, University of Washington
Fionn Murtagh, Royal Holloway, University of London

Proposals for the series should be sent directly to one of the series editors above, or submitted to:

Chapman & Hall/CRC
4th Floor, Albert House
1-4 Singer Street
London EC2A 4BQ
UK

Published Titles

Bayesian Artificial Intelligence
Kevin B. Korb and Ann E. Nicholson

Computational Statistics Handbook with MATLAB®, Second Edition
Wendy L. Martinez and Angel R. Martinez

Clustering for Data Mining:
A Data Recovery Approach
Boris Mirkin

Correspondence Analysis and Data Coding with Java and R
Fionn Murtagh

Design and Modeling for Computer Experiments
Kai-Tai Fang, Runze Li, and Agus Sudjianto

Exploratory Data Analysis with MATLAB®
Wendy L. Martinez and Angel R. Martinez

Interactive Graphics for Data Analysis:
Principles and Examples
Martin Theus and Simon Urbanek

Introduction to Machine Learning and Bioinformatics
Sushmita Mitra, Sujay Datta, Theodore Perkins, and George Michailidis

Pattern Recognition Algorithms for Data Mining
Sankar K. Pal and Pabitra Mitra

R Graphics
Paul Murrell

R Programming for Bioinformatics
Robert Gentleman

Semisupervised Learning for Computational Linguistics
Steven Abney

Statistical Computing with R
Maria L. Rizzo

Computer Science and Data Analysis Series

Interactive Graphics for Data Analysis

Principles and Examples

Martin Theus
Simon Urbanek

CRC Press
Taylor & Francis Group
Boca Raton London New York

CRC Press is an imprint of the
Taylor & Francis Group, an **informa** business

A CHAPMAN & HALL BOOK

Apple is a registered trademark of Apple Computer, Inc. AT&T is a registered trademark of AT&T Corp. DataDesk is a registered trademark of Data Description, Inc. SAS, SAS-Insight, SAS-JMP, and SAS-Stat-Studio are registered trademarks of SAS Institute Inc. and/or its affiliates. PostScript is a registered trademark of Adobe Systems Incorporated. SPLUS is a registered trademark of the Insightful Corporation. SPSS is a registered trademark of SPSS Inc. UNIX is a registered trademark of The Open Group. Windows is a registered trademark of Microsoft Corporation. Other third-party trademarks belong to their respective owners.

The figures on pages 157 and 213 are reproduced under the GNU Free Documentation License (GFDL) and can be found in the Wikimedia Commons. The figure on page 173 is copyright 2003 by Fenton, licensee BioMed Central Ltd., and is taken from the Open Access article "A new growth chart for preterm babies: Babson and Benda's chart updated with recent data and a new format."

First published 2009 by Chapman & hall/CRC Press

Published 2019 by CRC Press
Taylor & Francis Group
6000 Broken Sound Parkway NW, Suite 300
Boca Raton, FL 33487-2742

© 2009 by Taylor & Francis Group, LLC
CRC Press is an imprint of Taylor & Francis Group, an Informa business

First issued in paperback 2019

No claim to original U.S. Government works

ISBN-13: 978-0-367-45253-7 (pbk)
ISBN-13: 978-1-58488-594-8 (hbk)

Visit the Taylor & Francis Web site at
http://www.taylorandfrancis.com

and the CRC Press Web site at
http://www.crcpress.com

Library of Congress Cataloging-in-Publication Data

Theus, Martin.
    Interactive graphics for data analysis : principles and examples / Martin Theus, Simon Urbanek.
        p. cm. -- (Computer science and data analysis series)
    Includes bibliographical references and index.
    ISBN 978-1-58488-594-8 (hardcover : alk. paper)
    1. Graphical modeling (Statistics)--Data processing. 2. Statistics--Graphic methods. 3. Computer graphics. I. Urbanek, Simon. II. Title.

QA276.3.T54 2008
006.6--dc22   2008038130

To John W. Tukey

Preface

It is hard to resist starting this book with a phrase like "analyzing data is fun... ." Anyone who does a lot of data analysis knows that this is of course only partly true because the preparation of the data we want to analyze is often far less fun. Another limiting factor during a data analysis might also be the tools at hand. Interactive graphical tools address both these points — they are fun to use and efficient when it comes to understanding where you still have to clean up your data.

Most important, interactive graphical tools are the most powerful means for exploratory data analysis (EDA), which was postulated by John W. Tukey almost half a century ago. Being a visionary, Tukey talked about things he could only envision at that time. It took more than a decade before he could actually implement a first prototype. The PRIM-9 system belongs to history today, but what was important about Tukey's work was the philosophy of data analysis he brought into being. Some of his academic descendants carried this spirit on, and some even created new tools to support this new kind of data analysis. Paul Velleman's DataDesk^{TM} was the earliest benchmark for all of us.

To illustrate that it takes more than new tools to perform a more modern kind of data analysis, we want to quote an academic grandchild of John Tukey. Jay Emerson, student of John Hartigan, gives a short description of the course "Advanced Data Analysis" he taught at Yale:

> [...] I don't want you to leave the course feeling that you have learned about a limited set of tools, allowing you to do only certain types of analyses. I want you to feel prepared to face the unexpected, equipped with a set of skills enabling you to adapt to the inevitable surprises of data analysis. When faced with a fresh challenge, I want you to think, "I may not know the answer, but I bet I can figure it out." Someday, I want you to think, "that was one of the most practically useful courses I had at Yale."
>
> [...] You should be able to think critically about data, use graphical and numerical summaries, apply standard statistical inference procedures (when appropriate) and draw conclusions from such analyses. But most importantly, you should be willing to break out of the box and conduct new, innova-

tive analyses of problems when standard analyses may not be appropriate.

This course will be computationally intensive, and there is no substitute for getting your hands dirty. I expect to make my share of mistakes this semester (some intentional, some not), and we'll learn from them together. In data analysis, I believe you learn as much (and sometimes more) when things "don't work" than when they go as planned. You have succeeded when you can figure out why something doesn't work (or why some analysis isn't appropriate) and deduce an appropriate course of action as a result. You must be willing to try out new things and to make mistakes — you can't break the computer (at least, it won't go up in smoke), and the sky won't fall. Seek to understand the mistakes, and move onward.

Once you understand the differences and interactions between traditional statistics and data analysis, the usefulness of this book in learning and understanding data analysis by graphical means will be quite obvious.

It takes a radical skepticism against traditional statistical methods and a deep confidence into new approaches to move on to a new discipline. Having learned from and worked with Antony Unwin for many years, both of us started to appreciate this radicalism which allows us to distinguish between things more clearly.

We want to mention some of those who have had a sustained impact on our view of statistics and data analysis. To name just a few, we want to thank (in alphabetical order) Rick Becker, Axel Benner, Adrian Bowman, Andreas Buja, Dianne Cook, Jason Dykes, Fred Eicker, Michael Friendly, Wolfgang Härdle, Heike Hofmann, Kurt Hornik, Al Inselberg, Fritz Leisch, Junji Nakano, Balasubramanian Narasimhan (better known as Naras), Daryl Pregibon, Brian Ripley, Matt Schonlau, John Sammis, Günther Sawitzki, Robert Spence, Debby Swayne, Luke Tierney, Paul Velleman, Bill Venables, Chris Volinsky, Ed Wegman, Hadley Wickham, Rick Wicklin, Adi Wilhelm, Alan Wilks, Graham Wills and Achim Zeileis for their valuable inspirations.

We are also especially grateful to everybody who helped us find interesting and yet easy-to-understand and also easy-to-relate-to real-world datasets. These real life problems are vital input for the research and software development of interactive statistical graphics. The amount of time spent hunting for such datasets is often underestimated. Whereas the task in traditional parametric statistics is to find or simulate a (single) data set that works for a newly invented model class, the best way to move forward in interactive graphics research is to look for data that cannot be analyzed efficiently with the tools we have at hand so far.

Only a few of the datasets we have collected over the last decades were selected for this book. We especially want to thank Robert Erber for the collaboration on the medieval data on the city of Augsburg, the 110 students who took the Probability Exam in 2006, Antony Unwin for looking at the *Titanic* data from Dawson's *JSE* article in mosaic plots (shortly after initial skepticism on what I was doing with these funny plots), Rick Wicklin for sharing the data on the *Titanic* boats with us, Dianne Cook for pointing us to the Olive Oil data and the Tipping data and finally Annerose Zeis, who found a wonderful application of parallel coordinate plots with the Tour de France data (not to forget Sergej Potapov who made collecting the Tour de France data so much easier!).

We also want to thank Rob Calver for his considerate support at all stages of this book project and the reviewers who gave fruitful input that significantly improved the book.

Our final thanks go to our families for their infinite patience and understanding during the (too long) course of getting this book together. Without their support we would never have finished this project.

Martin Theus
Simon Urbanek

Contents

Introduction

What is this book about?

This book talks about exploratory data analysis (EDA) and how interactive graphical methods can help gain further insights in a dataset, generate new questions and hypotheses. John W. Tukey often referred to EDA as experimental work. Tukey and Wilk (1965) summarize data analysis saying it "... must be considered as an open-ended, highly interactive, iterative process, whose actual steps are segments of a stubbily branching, tree-like pattern of possible actions." Visualization of the data is probably one of the most powerful tools in this exploration process, as the role of the researcher in EDA is to explore the data in as many different ways as possible until a plausible "story" of the data emerges. Typical data analyses comprise the following eight steps:

1. **Plan the study**
   A well-thought-out study design that respects the study goals should be the initial step of any data analysis. For very clear-cut questions like optimizing the yield in a plant, optimal designs can be chosen — see Pukelsheim (2006), for instance.
   Unfortunately, statisticians are often consulted after the data were collected and thus cannot influence the study design. EDA methods can cope with the "here is the data" situation much more easily because they do not rely on apriori hypotheses and distributions.

2. **Understand the background and collect questions**
   Analyzing data without a further understanding of the background is almost impossible. This is often neglected in classical teaching of mathematical statistics. It is only if procedures and techniques relate to actual data that we might find interpretable results and be able to give proper recommendations. Thus, a study of the background and the data sources is extremely important for conducting a successful data analysis.

3. **Check the data for errors**
   From textbook examples, we are used to looking at what we regard as "clean" datasets. There are no obvious errors and the data seem to be consistent. But even for those datasets, the origin and the background of the data sometimes remains somewhat unclear and

a further exploration may reveal consistency problems or other oddities. An internet poll from KDD-Nugggets (http://www.kdnuggets.com/polls/2003/data_preparation.htm) shows that almost 2/3 of the respondents spend more than 60% of their time in a data mining project on data cleaning and data preparation. Obviously, typical data mining applications deal with complex mixtures of different generating processes. Even if we assume that the "classical" datasets we face in the daily business of a data analyst are of a better quality, there still remains a whole lot to do when initially checking the data for errors.

4. **Explore the data**
   Although any of the preceding steps might already be targeted toward a possible solution, the exploration of the data with the initially collected questions in mind is at the core of any data analysis. There is almost no limit regarding the possible data analytic tools which can be used at that stage. Sometimes we might want to collect more data or similar data with the same background. Other times we need purely computer-science-related techniques to get to the most interesting subset. In any case, we can benefit strongly from the use of (interactive) graphics and in order to be able to correctly judge what we see, we need an understanding of the fundamental concepts of randomness and statistical distributions.

5. **Review the initial questions**
   Sometimes checking the data for errors (step 3) might take us back to step 1, where the data were collected. Far more frequently, we might want to review the initially addressed questions after we gain further insight into the data during the exploration. New questions might arise and others might need to be reformulated or posed more precisely.

6. **Generate hypotheses and build statistical models**
   Statistical tests and models are used to separate signal from noise. When datasets are large, most of the effects we see graphically are actually significant; for smaller datasets, we need statistical tests to check for significance. With really large datasets almost every effect we might want to test will turn out to be significant.

7. **Analyze residuals and review hypotheses and models**
   Especially in a multivariate context, residuals of a statistical model could point to further structural features, which did not show up in any low-dimensional view. Be it a missing factor or outliers, or a further subsetting of the data, any remaining structure in the residuals might question the corresponding model.

8. **Interpretation and concluding recommendations**
   At the end of any analysis, we want to get to an interpretation of the results. This is tightly linked to step 2, where we should acquire a solid background understanding of the data. Interpreting the results means in particular to verify results for plausibility and check their relevance. Once this is done, we can usually give recommendations concerning the problem addressed with the data analysis.

Although these eight steps are written linearly, the whole process should iterate over and over again, until satisfactory results are achieved. For instance, we might be pointed to still undiscovered errors in the data only after we looked at the residuals of a statistical model. Interactive graphical methods are extremely powerful in steps three and four, i.e., checking the data quality and exploring the data. Judging the quality of a statistical model (step 7) can often be conducted more easily with suitable graphical methods at hand.

The book's structure is borrowed from Cox and Snell's (1981) book. About a quarter of a century after Cox and Snell wrote their book, statistics has gained substantially by utilizing computational and graphical methods and thus has changed. As many of the case studies in this book will show, the theory behind the traditional analyses is still valid. Interactive graphical methods complement the statistical toolbox to achieve more complete, easier to understand and easier to interpret analyses.

What requisites does it take to read the book?

The reader should have a basic understanding of statistical terms, and some knowledge about statistical models. A deeper insight into advanced statistical modelling and/or multivariate methods and methodology is helpful but not necessary to follow the ideas in this book as the book illustrates techniques from exploratory data analysis, rather than classical multivariate statistics or machine learning techniques. Nonetheless, the understanding and application of classical and modern multivariate techniques as well as machine learning techniques will strongly benefit from an effective graphical data exploration.

There are two icons that occasionally show up at the margins of the book. The first icon is the R-logo, which is shown whenever the book makes use of R-commands. We make use of R-code to quantify results more precisely, test statistical hypotheses or just rearrange data in a non-trivial way. If you are not familiar with R yet, you won't miss too much of the book, but be encouraged to start looking into R. The second icon is the "warning" road sign, which indicates that little more statistical or computing experience might be required to solve the exercise. Again, these exercises are intended as encouragement to dig deeper and they can be regarded as being optional.

Who should read this book?

This book should be enjoyable reading for anyone who has an interest in data analysis. As data analysis is the central application of all statistical theory — no matter from which discipline one looks at it — this book is certainly written for statisticians. Furthermore, anybody who needs to analyze and understand data, but has only a limited statistical education, will find this book useful. The examples come from various fields of application and are not limited to a specific area, such that every reader will find at least some examples he or she can directly relate to.

When used in a lecture, the book can be either taken as a stand-alone course on exploratory data analysis, if students already have taken classical introductory courses in statistics, or as a complement to a more traditional advanced (multivariate) statistics lecture. All chapters in the book feature exercises, which deepen and extend the content of the chapters. The book can be studied in a non-linear fashion, such as interweaving the principles and case studies. The first five main chapters in Part I cover most of the fundamentals of interactive statistical graphics and should be obligatory. Chapters 6–9 offer additional aspects and can be selected as necessary.

The book's website at http://www.interactivegraphics.org will give several suggestions on how to use the book in a course and offer additional material such as slides, extra code or exercises.

What is this book *not* about?

There are many very good books on statistical graphics. The first comprehensive work on quantitative graphical displays is probably Bertin's (1983) work, dating back to its first edition as early as 1967.[*] Around the same time much of John W. Tukey's work evolved, as summarized in Cleveland (1988). Whereas Bertin's work is full of inspiration, Tukey — being a statistician — always restricted himself to general applicable methods, a constraint many computer scientists in the field of data visualization too often forget about, even today. About 20 years ago, Cleveland (1985) continued Tukey's work by adding various perceptual considerations as well as the first experiences in dynamic graphics. Perceptual properties were the main focus of much of Edward Tufte's work (cf. Tufte, 1983 and Tufte, 1997) as well.

In the mid- to late 1980s, the desktop revolution in personal computing led to many projects in dynamic and interactive graphics (the difference will be discussed in Section 7.4). Cleveland and McGill (1988) give a collection of projects, all using the power of interacting with graphics to

---

[*]Note that the references mostly refer to the last significant edition, and not necessarily to the initial publication date

support statistical analyses. Although published almost 20 years later, the work by Young et al. (2006) can be seen as a direct follow-up, based on the XLispStat system (see Tierney, 1991, for details). As with all new technologies, the initial momentum slowed down, such that Cleveland's (1993) next book was more related to the visualization of statistical models again. Finally, Wilkinson (2005) put a lot of thought into the most careful construction of statistical graphics, setting up a formal notation to describe statistical graphics as generally as possible.

Wainer's (2004) book talks much about why and how a graphic can efficiently tell a story. Wainer nicely develops a story of the background problem and discusses the use of (static) graphics to better understand the underlying problem. The problems discussed are mostly of univariate nature and of a relatively small size. Much on the use of dynamic graphics in the context of multivariate statistics can also be learned from the book by Cook and Swayne (2007). Probably the most important early paper on interactive graphics is the work of Becker and Cleveland (1987). Further work, mainly within the computer science domain, often does not discriminate very well between the description of specific tools and general concepts related to data analysis. Nonetheless, concepts without the proof of a working prototype are often not very convincing. Two further references which primarily focus on data analysis aspects in interactive statistical graphics are Theus (1996) and Unwin et al. (1996).

This book is influenced by all of the above cited work; more by Tukey than by Bertin and more by Tufte than by Cleveland. It will not talk about diagnostic graphics for statistical models nor about presentation graphics. Both these aspects of statistical graphics have a very special purpose and usually do not contribute very much to an analysis. For the interplay of statistical models and diagnostic plots refer toVenables and Ripley (1999) andHeiberger and Holland (2004). The thoughtful construction of (not just) presentation graphics is discussed nicely in, e.g., Wallgren et al. (1996), Kosslyn (1994) and Robbins (2005).

At this point it is important to understand one fundamental difference between presentation graphics and graphics for data exploration. Presentation graphics face the challenge to depict a key message in — usually a single — graphic which needs to fit very many observers at a time, without the chance to give further explanations or context. Exploration graphics, in contrast, are mostly created and used only by a single researcher, who can use as many graphics as necessary to explore particular questions. In most cases none of these graphics alone gives a comprehensive answer to those questions, but must be seen as a whole in the context of the analysis.

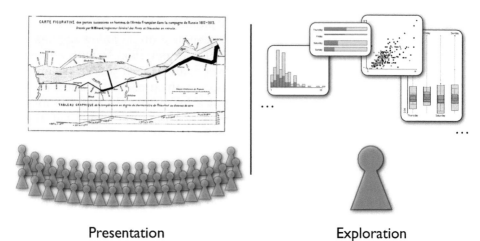

Presentation   Exploration

FIGURE 0.1

The relation between the number of observers and the number of graphics in use is inverse for presentation graphics and graphics for exploration.

Facing this distinction, it becomes quite clear that there is hardly a satisfying compromise[†] between the two aspects of the use of graphics. This also explains why most statistical software will not cover the two aspects equally well at the same time.

What is in the book?

Inspired by the book *Applied Statistics — Principles and Examples* by Cox and Snell (1981), this book is divided into two parts.

A persistent problem with data analysis is a lack of formalization. This has not changed ever since its inauguration. Tukey and Wilk (1965) see a clear need for formalization "... the technology of data analysis is still unsystematized ..." but also warn "Data analysis can gain much from formal statistics, but only if the connection is kept adequately loose."

Trying to keep the right balance between these two demands, the first part of this book summarizes principles and methodology, showing how the different graphical representations of variables of a dataset are effectively used in an interactive setting. Therefore, the most important plots are introduced along with their interactive controls. Different types of data or relations between variables need specific plots or plot ensem-

---

[†]Regarding compromises, professor Peter J. Huber once explained in a lecture at Augsburg University: "When a muckbucket is in your way, you can either pass it on the right, or on the left — who would ever consider a compromise... ."

bles to be examined properly. These plot ensembles are introduced and explained along with their counterparts in traditional statistics.

The second part of this book uses nine case studies to illustrate the principles introduced in the first part. Each case study consists of a description of the background, lists the main goals of the analysis and the variables in the data set, and shows what further — mostly numerical — procedures can add to the graphical analysis. Each case study ends with a summary of the most important findings and concluding remarks. Whenever an example has also been discussed in Cox and Snell's book, the numerical analysis for this dataset has often been taken from that book.

In the appendix, the software Mondrian (Theus, 2003) — which is used throughout the book — is described. All of the examples can easily be "rerun" by the reader using the software. Although the majority of the plots in this book are generated with Mondrian, they are usually plotted in an abstract form, in order to focus on the principles, rather than on a specific implementation in a tool. Only when certain user interactions must be explained, screenshots from a session are used to illustrate the interactive technique.

Both parts feature Exercise sections. These questions critically review the various chapter topics and address further issues not discussed explicitly in the chapters, rather than being a mere repetition of what the reader should have learned in the chapter. Unless mentioned otherwise, the software to use for these exercises is Mondrian. R (R Development Core Team, 2006) is the default choice for setting up simple statistical models or for performing non-trivial calculations which cannot be performed within Mondrian.

More on Software for Interactive Statistical Graphics

Almost every statistical software tool is able to draw graphics. These graphics are often static and usually optimized towards presentation quality graphics and thus not very helpful for an exploratory working style. Only when computers became personal and graphical, software tools for interactive graphics became accessible to more than only a few selected researchers.

One of the first programs, and still an impressive example of what can be done, is DataDesk (Velleman, 1997). DataDesk not only offers interactive statistical graphics, but also the ability to create models and their diagnostics in a comprehensive environment.

Xgobi (Swayne et al., 1998) — and its current successor ggobi (Swayne et al., 2003) — offer a wide support for dynamic graphics, i.e., Projection Pursuit and the Grand Tour (see Section 7.4 for further details).

Numerous software projects were initiated or hosted at the Department for Computational Statistics and Data Analysis at Augsburg University headed by Antony Unwin. MANET (Unwin et al., 1996) has a special focus on missing values, KLIMT (Urbanek, 2006) makes regression and classification trees alive and iplots (Urbanek and Theus, 2003), (Theus and Urbanek, 2004) finally adds interactive statistical graphics to R. Mondrian — which is used throughout this book — a stand-alone package with the broadest range of graphics and interactive features, was initially developed at AT&T Labs Research and largely extended in Augsburg. There is quite some similarity between iplots and Mondrian, such that many of the graphics used in this book can be created with iplots as well. However, the reader would need to learn the functions of the iplots packages as there is no simple point-and-click interface for creating plots as in Mondrian.

There are some more commercial packages to be mentioned here. John Sall's SAS JMP is an interactive environment quite similar to DataDesk, stronger in statistical models, though less equipped with interactive graphics. Both SAS Insight and its vastly improved successor SAS StatStudio are interactive graphics additions to the SAS system quite similarly to how iplots is an add-on to R.

Part I

Principles

1

Interactivity

Interactivity is key when using today's computers. Ever since the desktop revolution, we are used to manipulating graphical objects interactively. Today's computers are no longer abstract machines in the background, operated through punch-cards or command-line terminals, but have become personal workstations with a graphical user interface.

This chapter will introduce the essential concepts of interacting with statistical graphics. Querying statistical graphics to read off exact quantities, selecting subsets of interest in order to trace them in other graphics and modifying parameters of a graphic are at the core of interactive statistical graphics. These core concepts are explained in depth in this chapter and will then be used throughout all subsequent chapters without being explicitly mentioned.

1.1 Queries

To query a plot is probably one of the most intuitive interactions we can think of in interactive statistical graphics. Especially in a computer-based interactive environment it is a natural task to ask for information which is associated with an object in front of us. Querying information is often triggered by just pausing the mouse over an object, e.g., over

- an icon, we get the function of the icon displayed in a tooltip.

- a link in a web page will show the linked address either in a tooltip or in the status bar.

- an element of a graphic in a spread-sheet application will reveal the values of that object.

In interactive statistical graphics we are concerned about statistical objects, like points, lines or bars, which represent data. Querying such an object should show us the data represented by this object, be it a single observation or a whole group of observations.

Figure 1.1 gives two examples of queries for the *Olive Oils* dataset from case study H. In the left plot a barchart for the variable *Area* is queried,

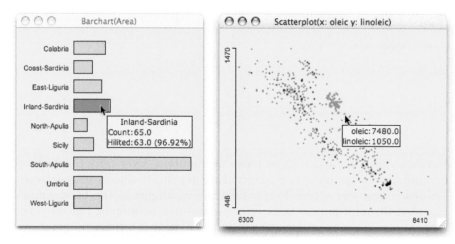

FIGURE 1.1

Querying a barchart gives the summary information for that category (*Inland Sardinia*) (left), whereas the query in the scatterplot (right) gives the coordinate values of the queried point.

i.e., a bar summarizing all cases for a particular category. The query shows the number of cases in this category (*Inland Sardinia*) and the number of highlighted cases in this category along with the percentage. We see that this group is selected almost completely except for two cases. The right plot of Figure 1.1 gives an example for a query of a single object, in this case a single point in a scatterplot. The query shows the values for the two variables in the plot for the queried point. Queries in Mondrian are triggered by pressing the <control>-key and moving the mouse over the object to be queried.

Generally there are two ways the query results can be displayed. In Figure 1.1 the query result is displayed contiguous to the object queried. An alternative is to display the queried information at a fixed place, for instance, in a separate query window. *Contiguous queries* have the advantage of showing the information in the context of the object, but may hide parts of the graphics. *Distant queries* do not cover the graphics, but the user may lose focus while reading the query results.

Levels of Queries

Queries can have different levels of detail. The Figure 1.1 example showed standard queries, which give the basic information on the queried object.

Typically, three levels of query detail can be distinguished. In an *orientation query*, the user may query the position of the mouse pointer in the coordinate system of the current plot. An orientation query can be

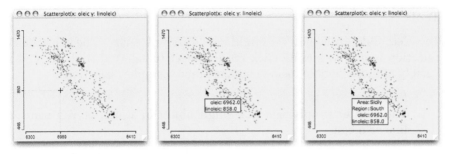

FIGURE 1.2
Example of three levels of queries, orientation (left), standard (middle) and extended (right) within Mondrian.

seen as the interactive equivalent to grid-lines typically used in static plots. The *default query* shows exactly the information associated with the object within the current plot. *Extended queries* can show any additional information which is linked to the object. Figure 1.2 shows an example of the three levels of query detail in a scatterplot. In the left plot an orientation query is shown which gives the coordinates of the mouse pointer. The middle plot shows the default query, which gives the exact values of the coordinates of the queried point. The extended query on the right shows not only the x and y coordinates of the point, but also the information on the variables *Area* and *Region* for this particular point. Extended queries can show far more than just numbers or text strings related to the query object, including pictures or even information from the web. Extended queries in Mondrian are triggered by adding <shift> to the default query.

1.2   Selection and Linked Highlighting

The most fundamental interactions in interactive statistical graphics are selections and the corresponding linked highlighting.

Selections

The first step of linked highlighting is to define a subset to be highlighted. Textual interfaces to select subsets of data are well known from query-languages for relational database systems like SQL. A graphical system needs graphical tools and techniques to define and select data.

Tools

To select data in a plot we might use different tools. The most basic selection tool is to simply click on an object to select it, but there is a variety of more flexible tools to support more complex selections:

- Pointer
  The pointer simply selects a single object in a plot, be it a point, a line or a bar.

- Drag-box
  The drag-box selects a rectangular region in a plot. Every object intersecting this rectangle will be selected.

- Brush
  The brush is a generalization of the drag-box. Once a rectangular region is defined, the brush allows us to move that region across a plot and thus dynamically change the selected subset.

- Slicer
  The slicer is a one-dimensional selection tool, which selects intervals along a single axis. Once an initial interval is defined, the upper or lower limit may be changed dynamically.

- Lasso
  All tools listed above rely on axis parallel selections. The lasso allows us to define an arbitrary contiguous shape to select data.

Obviously, not all software tools for interactive statistical graphics implement all selection tools, but the most common, i.e., pointer and drag-box, are usually supported. Furthermore, not all selection modes make sense in all plots. For instance the lasso might not be useful in a barchart, and the pointer less common to use in a parallel coordinate plot. Figure 1.3 shows two examples of selection tools. The left plot shows the selection of a cluster in a scatterplot via a drag-box. The right plot shows the use of a lasso. As the cluster highlighted in the right plot cannot be selected by a rectangle parallel to the coordinate axes, the lasso is used to circumscribe the points.

Modes

A basic one-step selection is usually sufficient for simple subsets. In some cases — especially for large datasets — more complex selections are necessary in order to define a subset of interest. To create multi-step selections, the following boolean operators can be used to modify an existing selection set:

- AND
  The AND-operator creates the intersection of the current selection

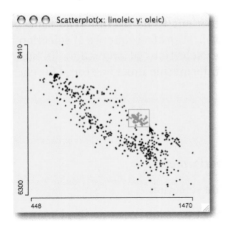
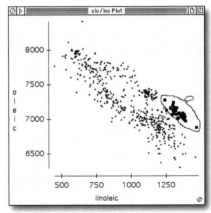

FIGURE 1.3

Two examples of selection tools: via a drag-box in Mondrian (left) and the lasso in DataDesk (right).

and newly selected data, i.e., only points that exist in both selections are included in the resulting subset.

- OR
  The OR-operator creates the union of the current selection and newly selected data, i.e., all points that exist in either one selection are included in the resulting subset.

- XOR
  The XOR-operator creates the exclusive union of the current selection and newly selected data, i.e., all points that exist in either selection, except for those that exist in both selections, are included in the resulting subset.

- NOT
  The NOT-operator (which is in fact a combination of NOT and AND) excludes the newly selected cases from the current selection.

A nice overview of the versatility of combining selections via boolean operators can be found in Wills (1996).

Selection Sequences

Composed selections are powerful to specify complex subsets, but once created, hard to maintain. After a selection operation has been applied to a current selection, this step cannot be undone or altered, nor can previous steps be changed, without redoing the complete sequence of selections performed so far.

At this point, *selection sequences* turn out to be useful. The concept behind selection sequences is to store each step of a sequence of selections, and thus to allow us to delete and alter selections at any stage. To build up a selection sequence, the following information must be stored for each selection:

1. Selection step

2. Selection mode

3. Plot in which the selection was performed

4. Screen and data coordinates of the selection

Figure 1.4 shows an example of selection sequences as implemented in Mondrian for the *Restaurant Tips* data (compare the case study in Appendix i). The subset which is selected is given by: "Find all customers, who paid less than 15% tip, at night, except on weekdays!" Each selection step is indicated by a selection rectangle, which can be altered at its 8 handles, or moved as a brush. The selection mode may be changed via the context menu of the selection rectangle and each selection can be deleted by simply hitting the backspace key.

Selection sequence can be seen as the graphical pendant to database queries. They can easily be translated into a database language like SQL.

At this point one has to note a property of selection sequences which is not directly obvious, but intuitive. In contrast to a query specified in, e.g., SQL, selection sequences are always explicitly left-parenthesized, e.g. a selection over the sets A, B and C with the corresponding WHERE-clause in SQL

```
      ...   WHERE A OR B AND C
```

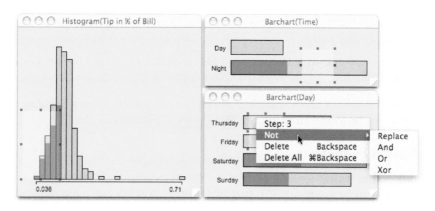

FIGURE 1.4
A selection sequence over three different plots.

or in mathematical notation

$$\text{Seq}_1 = A \cup B \cap C = A \cup (B \cap C)$$

is different from the selection sequence

$$\text{Seq}_2 = ((A \cup B) \cap C)$$

Wheras Seq_1 follows the usual bindings of boolean operators, Seq_2 is built up step by step from left to right — which is actually the way the user thinks of the selection which is built step by step.

Selection vs. Brushing vs. Painting

One further criterion to categorize a selection method is to distinguish between **transient** and **persistent** selection of cases. So far, a selection was always regarded as transient, i.e., the selection state of all cases was either the result of a single selection, or can be derived from the steps of a selection sequence.

Brushing, i.e., the dynamical movement of a selection region (usually a rectangular brush), can be either transient, or persistent. The process of persistent brushing is usually called **painting**. In painting mode, all objects which are selected with the brush are either selected — and stay selected — or deselected.

Technically, painting can be regarded as a one-step selection sequence, where the first selection step is set to OR-mode (or NOT-mode, if the brushed points should be deselected).

Persistent brushing can be used as a substitute for the lasso selection tool. The lasso is used to select arbitrary shapes, usually structures which cannot be captured by axis-parallel rectangles as produced by the drag-box or the brush. If the brush is small enough, painting can be used to select structures such as diagonals, curves or complex shapes.

Selecting, brushing and painting are conceptually quite different actions. Brushing is a process that continuously redefines a subset selection and needs an equally continuous observation of the selected subset during the brushing process. A one-step selection highlights properties of a single group of data across many views. Painting commonly reflects static group memberships which could possibly be produced in an static environment or are used in conjunction with dynamic graphics techniques as discussed in Section 7.4.

Highlighting

Once a selection has been performed the corresponding highlighting in other plots is of interest. For glyph*-based plots, e.g., scatterplots or parallel coordinate plots, the definition of how highlighting is performed is quite obvious: all highlighted cases shall be drawn in the highlighting color and sometimes even plotted in larger size. Figure 1.3 shows the two possibilities for a scatterplot. The left plot draws the highlighted points in red (in contrast to the black points, which are not selected), whereas the right plot — not relying on color — uses bigger symbols to indicate the selected points.

For plots that are not glyph based, the question of how to implement highlighting is a bit more complex. Two general questions arise:

1. Is the highlighting itself the same type of plot?
As an advocate of high consistency one might want to quickly answer this question with "yes." The following example illustrates that the answer is not that simple: Figure 1.5 shows a barchart for the variable *Day of Week* of the *Restaurant Tips* data. All cases where a smoker was present are selected. The leftmost plot is a traditional barchart, where the highlighting is again a barchart (it is actually a stacked barchart of selected and unselected cases). We clearly can see that the absolute amount of smokers is about the same for all days, except for Saturdays. On the other hand we have a hard time judging how the proportion of smokers differs from day to day. The rightmost plot in Figure 1.5 shows the same data for the weekdays in a spineplot. In a spineplot the length of the bars is no longer

---

*A glyph can be any kind of graphical object, but for most statistical plots we can assume it to be a point or a line.

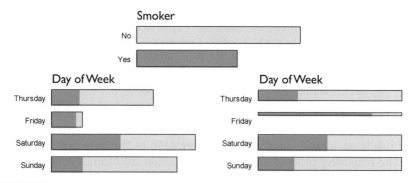

FIGURE 1.5
Depending on the question a plot should answer, the highlighting of a plot may not be of the same type as the plot itself.

Day of Week

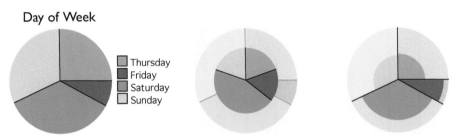

FIGURE 1.6
A piechart is not suitable for applying highlighting.

proportional to the number of observations in a class but set to a fixed unit length. The area of each bar is still proportional to the cell counts as the bar widths are adjusted accordingly. However, the highlighting is still performed in the same direction as in the barchart, enabling a direct comparison of the proportions of highlighting across the bars.

Both variants of the plot in Figure 1.5 have their benefits; the barchart allows the comparison of absolute numbers, whereas the spineplot makes it easier to compare the proportion of selected cases. Unfortunately not all graphics are suitable for highlighting and the principle we learned from the barchart example cannot necessarily be generalized to all plots. Figure 1.6 shows the same highlighting principles from Figure 1.5 applied to a piechart. The left plot shows the piechart for the unhighlighted data. The middle plot adheres to the principle of using the same plot for highlighting, whereas the rightmost plot uses a modification to allow a comparison of the proportions. Obviously none of the plots gives a good representation of the data and the selected subset, and comparisons across the different categories are quite hard.

The above question cannot be answered generally. In any case, the plot and its highlighting shall be easily or at least sensibly interpretable and answer the questions addressed by this plot.

2. Should the highlighting compare to the whole sample or its complement?
Again, the question can trivially be answered for glyph based plots, as these plots compare the selected cases with both the complement and the complete sample depending on how we interpret the plot. Even plots that summarize data like barcharts and spinplots do not really need this distinction.

Problems arise whenever more complex statistics are incorporated in the plot. One of the most common plots where we can illustrate this problem is the boxplot. Figure 1.7 shows two examples of the difference that the two alternatives can make. The left plot shows the alternatives for

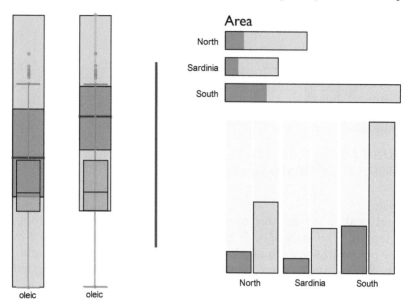

FIGURE 1.7
Highlighting may compare against the whole sample or the complement of the selection. Left, for boxplots; right, for barcharts.

boxplots. The left boxplot shows the complete sample in the gray boxplot[†] and the selected sample superposed. The right boxplot compares to the complement of the highlighted subset. The gray boxplot of the complement is quite different to the boxplot of the whole sample. The right plot in Figure 1.7 shows two versions of barcharts. The upper barchart shows the traditional highlighting, whereas the lower barchart shows two barcharts side by side, one for the selected subset and one for the complement.

Usually a selected subgroup is relatively transient and thus the corresponding highlighting. Many different subgroups of interest may be visited during an exploration and thus the highlighting may change very often. The reference to compare against should then not depend on the current selection, which makes the comparison against the whole sample preferable. The comparison against the complement can also be generalized from a binary distinction between selected group and unselected group to a discrimination of many groups, as will be shown in the following section.

[†]Note that this boxplot has been modified to allow a highlighted boxplot to be superposed — the whiskers are light gray boxes now; see also Section 2.2.

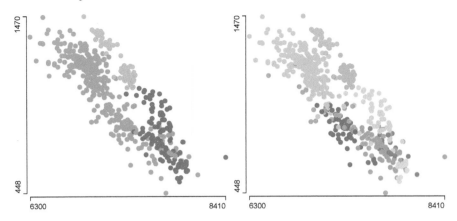

FIGURE 1.8

Two examples where color brushing works quite well. The plot on the left discriminates the three areas from the data in case study H, the right plot the nine regions.

Highlighting vs. Color Brushing

Color brushing is the technique that assigns a color to each observation. Usually the color is defined by a grouping variable of some mutually exclusive classification scheme. In contrast to selection and highlighting, this assignment is persistent, whether we use optional highlighting or not. Color brushing is well defined for glyph-based plots, as each glyph uses the color that is assigned to the corresponding case. For plots showing summaries of the data, a discrimination of more than two groups is usually hard to implement.

Color brushing can be very effective in cases where more than just a specific subgroup should be highlighted. Figure 1.8 shows two scatterplots where color brushing works quite well, as the number of classes to distinguish is either small, or the groups separate well.

In Figure 1.8 we already see one of the problems which arise from color brushing. The z-order, i.e., the stacking order of the points, is easy to handle if only two groups (e.g., highlighted vs. other cases) have to be distinguished, because highlighted points are always plotted on top. Having more than two groups, there is no longer a natural z-order defined. For an illustration, we look at the two parallel coordinate plots in Figure 1.9.

The plot shows the fatty acid content of 572 Italian olive oil samples (see case study H) in a parallel coordinate plot. In both plots the lines are colored according to the area the sample was taken from. There are only three areas (*North*, *Sardinia* and *South*) to distinguish. In the upper plot, the lines are plotted in the order the cases appear in the dataset (usually it is hard to find an order in a dataset which is generally meaningful other

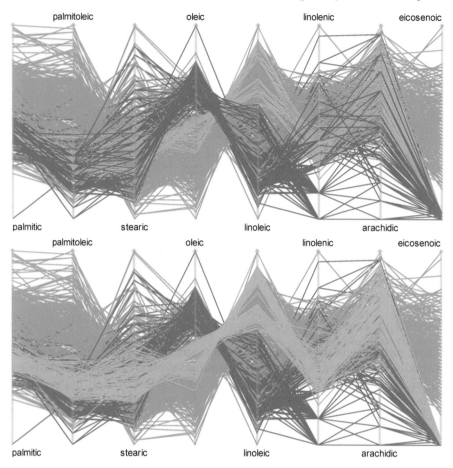

FIGURE 1.9
The upper parallel coordinate plot shows the cases in the order of occurrence in the dataset. The plot below shows the group of *Sardinian* oils explicitly on top of all cases. Note the strong difference in the visual impact of the two plots.

than the order provided by the dataset itself). At a first sight it does not seem that the visual impact of the plot is impeded by overplotting very much. The lower plot shows the same plot, except for the fact that the purple colored group of *Sardinian* oils is now plotted on top of all other cases. The visual impact is quite different. Although we are facing only 3 groups, there are 6 potential stacking orders for these 3 groups. In general, it takes $k!$ plots to look at, if we have k colored groups. If we were to look at the 9 regions in the dataset, it would take 9!=362,880 potential plots. Certainly, not all of them really reveal different features, but those that do so are still too many to look at.

The problem of overplotting can be as severe that (smaller) groups can disappear completely, which will not only lead to quantitatively biased inferences, but even to qualitatively inappropriate conclusions.

Other problems with color brushing arise when used alongside with highlighting. Obviously the highlighting color and the colors used for brushing may interfere. Although the perception, assignment and preference of colors is very subjective[‡] these problems can easily be taken care of. The more difficult problem to solve when using highlighting and color brushing simultaneously is masking. Both color assignments are usually completely independent categorizations. That means that if k colors are assigned by color brushing and another binary classification of the data is present by means of highlighting, we get a classification of size $k \times 2$. Unfortunately, the colors assigned by color brushing are masked by the highlighting color for all selected cases.

Figure 1.10 shows an example of color brushing and highlighting used simultaneously. The data are color brushed according to *Day of Week*, and all females are highlighted in red. This would result in $4 \times 2 = 8$ groups

---

[‡]See www.colorbrewer.org for an excellent compilation of color schemes for different purposes.

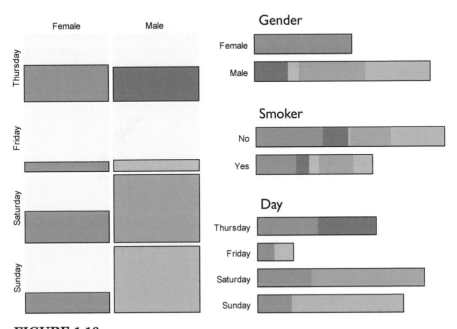

FIGURE 1.10
Color brushing and highlighting (in red) used simultaneously.

which need to be distinguished, but only 5 groups are actually visible, i.e., all highlighted cases in red, unselected Thursdays in blue, ..., unselected Sundays in orange. These 8 groups are the 8 classes visible in the mosaic plot on the left of Figure 1.10. Note that in the barchart for smoker, each bar is divided into 5 groups, selected cases and unselected cases according to *Day of Week*.

Apart from color brushing, where all cases of a dataset get a color assigned, there is also the possibility of assigning just one or a few colors to specific groups like outliers or sample errors. Such a persistent assignment produces less optical clutter and can be helpful as a constant reminder of "special cases." Nonetheless, this procedure has its limits as well. Unlike color brushing by the mutually exclusive classes of a categorical variable, individually assigned groups may well intersect. The obvious problem arising then is how to color the intersection of two color assignments.

Hot-Selection and Shadowing

There are two more interactive techniques which are closely related to selection and highlighting. The so-called *hot-selection* mode is a property of a single plot and simply means that only currently selected cases are shown in a plot. Whenever the selection changes, the plot will change as well, thus allowing us to focus on the actually selected subgroup. Hot-selection comes in two flavors, either with fixed (respectively frozen) scale, or with automatic scale, i.e., the scale is updated whenever the selection changes. Shadowing — which is sometimes also referred to as *ghosting* — hides points in order to focus on the rest of the sample. In contrast to hot-

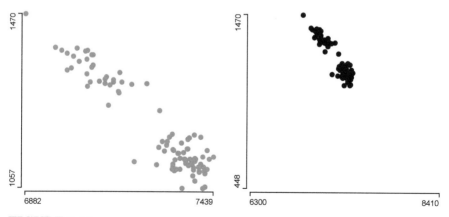

FIGURE 1.11

Left: Hot-selection with automatic update of the plot scale, right: ghosting. Both plots focus on the group of Sardinian oils from Figure 1.8 (left).

selection, shadowing is a persistent feature of the data, and usually does not affect the plot scales. Ghosting can also be applied with not only two states (on/off), but may also be used to hide objects just to a certain degree of $(1 - \alpha)\%$. This is directly related to the technique of α-transparency, which is discussed in Section 3.3 and Chapter 9 more deeply.

Figure 1.11 gives an example each for hot-selection and shadowing using the scatterplot shown in Figure 1.8 left. The left plot in Figure 1.11 shows the hot-selection scatterplot for the Sardinian cases with an updated scale for this particular selection, whereas the right plot shows the effect of shadowing for the same scatterplot.

1.3 Linking Analyses

So far, the effect of a selection was restricted to the corresponding highlighting in linked graphics. Linking is not limited to graphics, but can also be applied to statistical summaries, tests or even models. Implicitly, we find linked statistical summaries in a graphical form in every barchart and boxplot. The barchart dynamically displays the result of the breakdown summary for the selected subset of a categorical variable; the boxplot shows the five-number-summary of the selection of a variable. In Figure 1.12 a regression analysis of the currently selected cases is shown along with the corresponding scatterplot and regression line. Whenever the selection changes, the results of the regression analysis are updated. The regression summary is set to hot-selection mode and reflects the properties of the highlighted regression line.

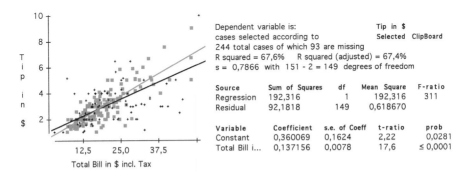

FIGURE 1.12

A hot-selection regression analysis in DataDesk along with the corresponding scatterplot and regression line. Whenever the selection changes, the regression summary will update immediately.

Although very powerful, most statistical software does not match the interactivity needed to perform statistical analysis on dynamically changing subsets. The example in Figure 1.12 is taken from a DataDesk session.

Obviously, statistical tests and models are not necessarily part of interactive statistical graphics software. Nonetheless, graphical software may incorporate the results in graphical form, such as regression lines, scatterplot smoothers or density estimates in order to enhance the statistical plots.

1.4   Interacting with Graphics

Selection and linked highlighting are only the most general and fundamental aspects of interacting with graphics. There are further interactions that can be applied to most statistical graphics like zooming and sorting, explained in more detail in Chapter 7.

Apart from general interactions, various plot-specific interactions exist. These interactions can be used to change plot-specific parameters. How quickly these changes can be performed depends strongly on the user interface chosen. Figure 1.13 gives a brief ranking of the most common user interface controls according to how much context they preserve and the amount of information they can carry. Plot-specific interactions can

FIGURE 1.13

Different user interface controls offer different possibilities. Context preservation and speed are negatively correlated with the amount of information and complexity a command can hold.

be for instance:

- Setting the anchor point and bin width of a histogram (bandwidth of a density estimate)

- Changing the point size in a scatterplot

- Flipping the axes in a scatterplot

- Setting the smoothness of a scatterplot smoother

- Switching the representation (relative or absolute) of a barchart/spineplot or histogram/spinogram

- Changing the amount of α-blending applied to a glyph based plot (see Chapter 9 for further information on α-blending).

Although one could think of further interactions, the list is still relatively short. This is because many interactions are universal to all plots and not bound to a specific plot. All interaction like:

- Creating, and manipulating selections

- Changing the order of objects

- Changing scales (zooming)

are available in almost all plots and thus can be discussed in a more general way. These general interactions must share the same user interface controls across all plots and can be implemented consistently. This makes them easily accessible and flattens the learning curve of the software drastically.

Exercises

1.1. Queries
Test all query levels in all plots in Mondrian. Where do you find differences, where similarities?
Are all levels defined in all plots? If not, can you find good reasons for the differences?

1.2. Selections

(a) Which boolean operation does pressing the <shift>-key add to a selection?

    (b) Write down in formal notation: how can AND be represented with OR and XOR?

    (c) Which are the minimal sets of boolean operators, such that AND, OR, XOR and NOT can all be expressed?

1.3. Selections

Select the same data as selected in Figure 1.4 *without* using selection sequence! How can a <alt>-<shift>-selection help?

1.4. Selection Sequences

Create the selection sequence from Figure 1.4. Redefine the tipping rate to be at least 20%. Could a different sequence define the same set?

1.5. Color Brushing

For the rental data from case study E

    (a) Color the scatterplot of *Rent* vs. *Area* according to the number of rooms.

    (b) Color brush the data according to the year of construction. What can be learned from the barchart of *Number of rooms*?

2

Examining a Single Variable

This chapter introduces the most important statistical graphics to investigate the structure of a single variable. The plots in this chapter can be best categorized by the scale of the variable they display — either continuous or categorical. The latter includes both alphanumerical (nominal) and numerical (possibly ordinal) categorical data.

The most important interactive controls and variations are explained for each plot. Furthermore, the definition and use of highlighting are described.

2.1 Categorical Data

Barcharts and Spineplots

One of the most basic plots is the barchart. As the name implies, a bar is drawn for each category of the variable. The length of each bar is proportional to the number of cases falling into that particular category.

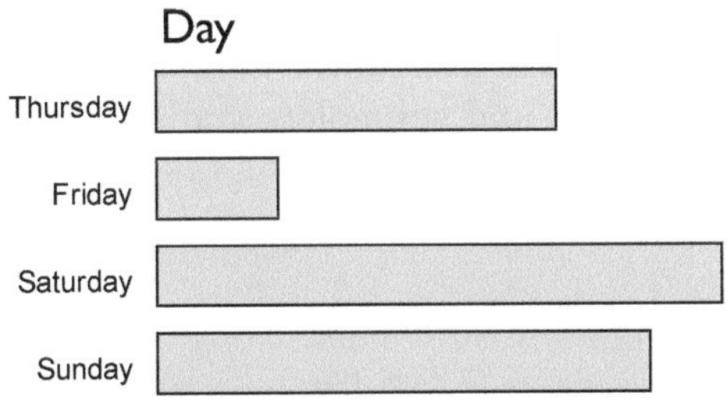

FIGURE 2.1

A barchart for the four days of the week from the Tipping data case study.

Barcharts can be either drawn vertically (what most applications do) or horizontally. Although the vertical layout is probably the more natural way to plot bars, the horizontal layout has the advantage of allowing bar labels to be printed in full length. This is especially useful when working with many categories. Figure 2.1 shows a barchart in a horizontal layout for the variable *Weekday* from the case study in Appendix i, the *Tipping* data. Note that the order of the bars matters a lot in this example. The default order — usually a lexicographic order — would place *Thursday* last, which would make a correct interpretation of the plot unnecessarily complicated.

Interactive Graphics for Data Analysis

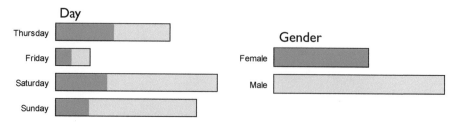

FIGURE 2.2

Highlighting in a barchart.

Adding highlighting to barcharts is straightforward. The barchart of the selected data points is simply drawn on top of the base barchart of the whole sample. Figure 2.2 gives an example of a highlighting in a barchart. All female customers have been selected in the same barchart as shown in Figure 2.1. We see immediately that the distribution of the highlighted cases is not of a different structure than the whole sample, but still, the proportion of females appears to be larger on Thursdays than on Sundays. What makes the comparison of the proportions so difficult? The highlighted part of a bar must be normalized in order to be comparable — a visually challenging task.

Spineplots use normalized bar lengths while the bar widths are proportional to the number of cases in the category. Figure 2.3 shows the data from Figure 2.2, but now the barchart has been switched to a spineplot. The area of the bars is still proportional to the category frequencies. The highlighting proportion can now be compared across all categories since the highlighting direction remains unchanged relative to the barchart and the area of the highlighting is also proportional to the highlighting frequencies. Now we see directly that the proportion of female customers declines monotonously from more than 50% to less than 25% from Thursdays to Sundays. In Mondrian, barcharts and spineplots use the same framework and switching between the two representations takes pressing only a single key.

FIGURE 2.3

The same data as in Figure 2.2 in a spineplot.

Looking at absolute and relative amounts of highlighting shows the
need for sorting and reordering of barcharts, which is discussed in Chapter 7. The generalization of barcharts and spineplots for more than just
one variable, i.e., the mosaic plot, is introduced in Chapter 4.

2.2 Continuous Data

There are far more plots for continuous data than for categorical data.
Obviously the amount of information a continuous variable can hold is
far greater than a categorical does. Depending on what aspect of a continuous variable is of interest, the one or the other graphic might be the
better choice.

Dotplots

Dotplots are a very simple way to plot one-dimensional data. Nevertheless, there are at least three distinct versions. The **standard dotplot** is
a scatterplot in one dimension, i.e., a continuous variable is plotted along
one axis only. Figure 2.4 shows an example of a standard dotplot for the
variable *Tip in USD*. Although this dataset has only 244 observations, we
note a strong overplotting for smaller values. No structure is visible for
tips less than $4, whereas the outliers beyond $6 can be easily spotted.
    Jittering is often applied to avoid overplotting in glyph based plots.
Jittering is a technique where a small amount of noise — usually white
noise, i.e., uniformly distributed random numbers — is added to the data
to avoid overplotting. Figure 2.5 shows the same dotplot as in Figure 2.4
now with a small amount of noise added orthogonally to the x axis. In
the **jittered dotplot** far more structure is visible in the data even for
amounts of less than $4. The jittering reveals accumulation points at
amounts of $2.00, $2.50, $3.00 and $4.00. Obviously, many customers
tend to give a tip of multiples of half a dollar.

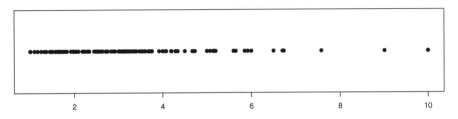

FIGURE 2.4
A standard dotplot for the variable *Tip in USD* of the *Tipping* data.

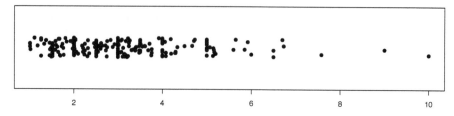

FIGURE 2.5
The same data as in Figure 2.4 with jittering applied.

Although jittering reduces the amount of overplotting considerably, it also introduces a negative effect. The pseudo structure along the virtual y axis may add visual artifacts, as the reader of the graphics will be inclined to interpret the location of points along the y axis as well. A first step to reduce this negative effect is to use so-called **textured dotplots** as introduced by Tukey and Tukey (1990). Textured dotplots use a systematic way to place points side by side to avoid overplotting. There are two drawbacks of such an approach: first, it is impossible to seamlessly switch between the systematic placement and a random placement of points, which is needed for larger datasets. Second, a systematic placement of points needs to have some idea of the density of the variable displayed, which is already a far more complex concept than the initially simple concept of a dotplot. Figure 2.6 shows a dotplot where the amount of jittering is proportional to the data density. This reduces potential visual artifacts due to pseudo structure along the y axis. This variation of a dotplot is probably a data representation giving best insight into the structure of the variable, but also far more complex than a standard dotplot.

A nice overview of the generation of many variations of dotplots can be found in Wilkinson (1999). In an interactive context, the amount of jittering should be controlled interactively and the added noise should be resampled when requested by the user.

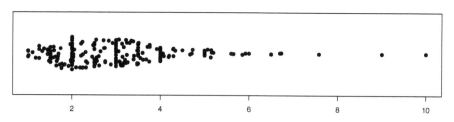

FIGURE 2.6
A dotplot with jitter added which is proportional to the data density.

Boxplots

The boxplot is a graphic which depicts both summary statistics as well as raw data. At the core of a boxplot is the so-called *five number summary*. The five number summary of a variable consists of the minimum, lower hinge, median, upper hinge and maximum. The definitions of the median and the extreme values are well known. The hinges are the medians of the subsamples which are created when dividing the original sample into two parts at the median. Thus, quartiles (0.25 and 0.75 quantiles) and hinges may differ by one index in the sorted sample, which usually does not change the resulting boxplot. Therefore the hinges are often substituted by quartiles for simplicity. Figure 2.7 illustrates all components of a boxplot. The core — the box — is built up by the upper and lower hinges and the median. The difference between the hinges — the so-called *h-spread* — is used to define the *inner fence* and the *outer fence.** The *whiskers* are drawn from the upper (resp. the lower) hinge to the first value which is no further away from the hinge than 1.5 times the h-spread — the inner fence. All points between inner and outer fence are called *outliers*; all points further away then 3 times the h-spread (the

---

*The inner and outer fences are never drawn in a boxplot, as they are only imaginary thresholds.

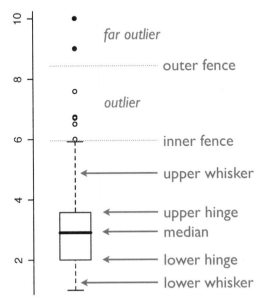

FIGURE 2.7
A boxplot with all its elements annotated.

outer fence) are called *far outliers*. Far outliers are usually marked with a more distinct symbol.

Comparing the boxplot in Figure 2.7 with the three dotplots above shows advantages and disadvantages of this display. The boxplot shows robust measures of location and spread, which gives basic properties of the sample's distribution. These properties are impossible to determine from a dotplot. On the other hand, all data points which are not outliers are represented in an abstract way. Thus it is impossible to see gaps or accumulations in a boxplot, both of which are easy to spot in a dotplot.

Highlighting in a boxplot must respect its special structure made up by summaries and single values. Whereas it is obvious that the glyph of a selected outlier can be highlighted easily, it is not sensible to highlight the box of a boxplot the same way the box in a barchart is highlighted. An example of how highlighting in boxplots can be implemented is shown in Figure 2.8. The upper boxplot shows a base boxplot without any highlighting. In order to be able to plot a highlighted boxplot atop the base plot, the whiskers have been modified and are now light gray boxes extending the inner box of the boxplot. The lower boxplot shows the highlighting. A regular boxplot for the highlighted cases is plotted in the highlighting color atop the base plot. The box of the highlighted boxplot is narrower and slightly transparent such that the parameters of the base boxplot are not obscured.

The definition of a boxplot has some desirable properties, in particular when we assume the data to follow a normal distribution. 50% of the data around the data center lie in the box — regardless of the distribution. For a (standard) normal distribution the quartiles can be found at -0.675 and 0.675 so that the h-spread is approximately 1.35. Adding 1.5 times to the box yields an interval of $[-2.698; 2.698]$. The probability that we observe values outside this interval is $P(x \notin [-2.698; 2.698]) = 0.7\%$. Thus the

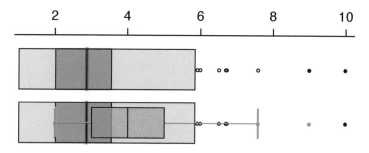

FIGURE 2.8
Highlighting in a boxplot needs a modification of the rendering of the unhighlighted boxplot.

probability that a value is an outlier is just below 1%.[†]

Histograms and Spinograms

Histograms are based on a summary of the sample. For each interval in a set of consecutive intervals, the number of observations falling into that interval is counted. The resulting counts are visualized with bars plotted over the intervals. The area of each bar is proportional to the corresponding count for this interval. The intervals — which are called "bins" — are usually set to have equal width and to be left closed and right open. Figure 2.9 gives an example of two histograms. Both histograms show

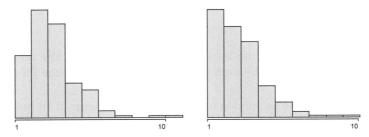

FIGURE 2.9
Two histograms of the variable *Tip in USD*. The left histogram uses bin width of exactly $1, whereas the right histogram uses slightly wider bins of width $1.01. Both start at $1.

the same data. The left histogram uses interval bin width of exactly $1, the right histogram a slightly larger bin width of $1.01. Both histograms start at $1. The apparent shape of the distributions looks quite different. As we see from this example, the parameters that determine a histogram are *bin with* and *anchor point*. For the data in Figure 2.9 a bin width of $1 seems more justifiable, and thus the resulting histogram is probably the better choice.

The dotplots showed accumulations at full and half dollar amounts, which are not visible in either of the histograms of Figure 2.9. To find these accumulations, the bin width has to be set to even smaller amounts than $1.00. Figure 2.10 shows the same data for bin widths $1.00, $0.50 and $0.25. The smaller the bin width, the more apparent are the accumulations at full dollar amounts of $2.00 and $3.00. For a better comparison, the scales of all three histograms have been set to be equal. Since the

---

[†]An anecdote says that John W. Tukey answered the question why it is 1.5 times the h-spread with: "Because 1 is too small and 2 is too large."

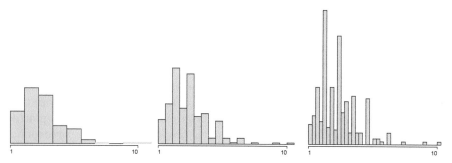

FIGURE 2.10

Three histograms of the variable *Tip in USD*. The chosen bin widths (from left to right) are $1.00, $0.50 and $0.25. All histograms share the same scale and start at $1.

area of bars is proportional to counts (or proportional to the relative frequency), the sum over the area of all rectangles is the same for all three histograms.

It is obvious from examples in Figure 2.9 and Figure 2.10 how important it is to be able to change bin width and anchor point of a histogram quickly and flexibly. Interactive controls of a histogram must allow us to change the two parameters by a simple mouse drag or keyboard shortcuts. If changing the parameters involves retyping a command and/or creating a whole new plot, the analyst might be inclined to avoid looking at many different views.

Highlighting in histograms can be implemented easily. A highlighted histogram of the selected cases is drawn atop the histogram of all cases. In Figure 2.11 a histogram of the rental price per area is plotted for the data from case study E. All apartments classified to be located in a "good" neighborhood are selected. The immediate question which arises is whether the distribution of the highlighted cases is any different from the distribution of all cases. This question is difficult to answer from the highlighted histogram in Figure 2.11 since we would need to compare the proportions of the highlighted cases across all bars of the histogram, which is visually unfeasible.

One way out is to use the same "trick" as switching from barcharts to spineplots, i.e., all bars are normalized to have the same height, but proportional width. The resulting plot is called a **spinogram** (cf. Hofmann and Theus, under revision). Spineplots have the nice property that highlighted proportions can be compared directly. However, it must be noted that the x axis in a spinogram is no longer linear. It is only piecewise linear within the bars. Although this might be confusing at first sight, it yields two interesting characteristics. Areas where only very few cases have been observed are squeezed together and thus get less visual weight.

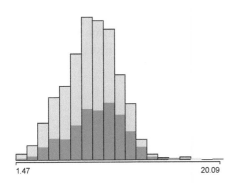 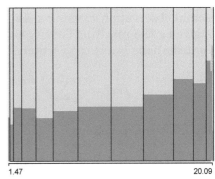

FIGURE 2.11
The variable *Rent per m²* from case study E. Left: a histogram with all apartments in a "good" neighborhood highlighted, right: the same data in a spinogram.

Let $\tilde{F}$ denote the empirical distribution function of a variable X, then the x axis in a spinogram is linear in $\tilde{F}^{-1}$. Applications and extensions of spinograms will be further discussed in Section 3.2.

As both views, histograms and spinograms, offer specific insights, it is desirable to switch quickly between the one and the other view, without being forced to create a new plot.

Density Estimation

Histograms are often used to visualize the density of a one-dimensional continuous sample. Figure 2.9 and Figure 2.10 illustrated the strong variation of histograms with the change of the bin width and the anchor point. Histograms are powerful in cases where meaningful class breaks can be defined and classes are used to select intervals and groups in the data. However, they often perform poorly when it comes to the visualization of a distribution.

This drawback was identified long ago, and several strategies have been taken to overcome this problem. One solution is to use so-called **average shifted histograms** or ASHs for short (see Scott, 1992). The idea behind average shifted histograms is quite simple. For a given bin width, the anchor point of a histogram can be shifted within the range of one bin width. Using k different starting points will result in k different histograms. For any given x the average over the k bar heights can then be computed to construct a smoother estimate of the underlying density.

Another method to visualize the density of a variable is to use **kernel density estimators**. The idea behind kernel density estimators is as follows. Given a sample of the size n, each observation contributes $1/n$-th of the density. This contribution to the density is distributed around the actual observation x_i using a scaled kernel function $k(x)$ at point x_i. For a given x the resulting density estimate can then be summed up over all contributions $k_{x_i}(x)$ each centered around the n x_i's, yielding

$$\hat{f}(x) = \frac{1}{cn} \sum_{i=1}^{n} k\left(\frac{x - x_i}{c}\right) \quad \text{for} \quad k(x) = k(-x). \tag{2.1}$$

For all kernel functions k

$$\int_{-\infty}^{\infty} k(x)dx = 1, \quad \int_{-\infty}^{\infty} k^2(x)dx < \infty, \quad \left|\frac{k(x)}{x}\right| \to 0 \quad \text{for } |x| \to \infty. \tag{2.2}$$

Figure 2.12 illustrates how a kernel density estimate is assembled from n kernel functions for the x_i using a normal density as kernel. The $n = 244$ cases of the variable *Tip in USD* with their corresponding kernels $k(x_i)$ are plotted in blue. Summing these functions up for each x gives the resulting density estimate plotted in purple. Note that the functions of the blue and the purple curves are drawn on a different scale such that the kernel functions are visible. Various kernels can be used such as rectangular, triangular or normal.

It can be shown that ASHs converge for $k \to \infty$ toward a kernel density estimate with a triangular kernel (see Venables and Ripley, 1999).

Figure 2.13 shows the three histograms from Figure 2.10. Each histogram has a kernel density estimate superposed which uses a bandwidth c equal to the bin width of the underlying histogram. The leftmost estimate is clearly oversmoothed and cannot capture the structure of the variable, whereas the rightmost estimate looks quite rough and is thus

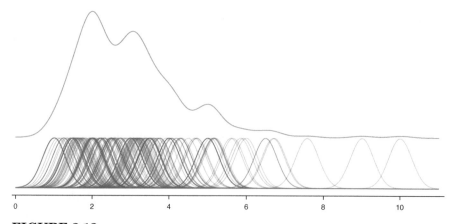

FIGURE 2.12
Illustration of how a kernel density is assembled out of the n contributing points x_i.

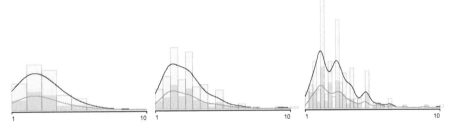

FIGURE 2.13
The same histograms as in Figure 2.10 now with kernel density estimates superposed. Male smoker parties are selected.

not satisfying. This trade-off — also generally referred to as bias-variance trade-off — can be best investigated when the bandwidth c of the density estimator can be varied interactively such that the analyst can see the change of the estimate instantaneously. For the data in Figure 2.13 there seems to be no 'best' bandwidth which captures the composite density.

CD-Plot

Although the spinogram is an efficient, area proportional display to visualize the conditional distribution of a subgroup of a continuous variable, the transformed x-axis of the spinogram can be difficult to interpret in some cases. The CD-plot visualizes the conditional distribution (CD) by

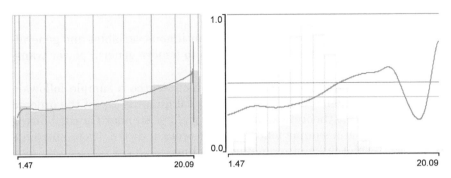

FIGURE 2.14
The variable *Rent per m^2* from case study E. Left: a spinogram with all apartments in a "good" neighborhood highlighted and a density estimate superimposed, right: the same data in a CD-plot. The CD-plot preserves the scale, whereas the spinogram focuses on intervals with a significant signal.

setting the density estimate of the selected subgroup in relation to the density estimate of the complete sample. The histogram is used just as a backdrop for orientation.

Figure 2.14 shows a CD-plot for the same data as in Figure 2.11. The trend in the data is the same as for the spinogram. The strong variation of the estimate for prices between €15 and €20 is due to the small number of points on which the estimate is based in this interval. This statistically insignificant information is avoided in spinograms since areas of very low density are squeezed to intervals of almost zero size.

2.3   Transforming Data

Continuous Data

Transforming data can have many motivations. A method requires normally distributed data to perform correctly, an extreme skewness of a distribution squeezes 99% of a variable's data onto 1% of the range of the variable, or the data simply needs a transformation into an established, more readily interpretable scale.

Transforming a variable is one of the earliest procedures found in dynamic graphics. The Box-Cox transformation defined by

$$x_{BC}(\lambda, \alpha) = \begin{cases} \dfrac{(x+\alpha)^\lambda - 1}{\lambda} & \text{for } \lambda \neq 0 \\[2ex] \ln(x+\alpha) & \text{for } \lambda = 0 \end{cases} \tag{2.3}$$

is the most common transformation for continuous variables and generalizes a simple logarithmic transformation to a more general power transformation.

A **qqplot** is often used to verify to which degree a sample follows a specific distribution. They plot the empirical quantile $x_{(i)}$ of the ordered sample against its theoretical quantile, e.g., for a standard normal distribution $z(\frac{i}{n+1})$. If a sample follows the theoretical distribution, all points in the scatterplot fall approximately on a line.

Figure 2.15 shows a qqplot for the variable *Tiprate* from case study i. Obviously the data are not normal, as several points deviate strongly from the line.

There are several approaches to make the data more normal. The simplest — and often used — is to take logs. In many situations the technical background even allows us to interpret the logarithm of a quantity (e.g., the quantification of sound waves).

The resulting qqplot is shown in Figure 2.16 left. As the result is not satisfying, a Box-Cox transformation with an optimized λ-value might be more appropriate. Using a dynamic transformation in DataDesk it is easy to find that for $\lambda = 0.155$, the variable is no longer skewed and the shape is very close to a normal distribution (λ has actually been chosen to match the skewness $\gamma_1 = 0$ and a kurtosis $\beta_2 = 3$ of a standard normal distribution). The resulting qqplot in Figure 2.16 (center) shows an improvement over the plain log-transformation, but many points still deviate from the line.

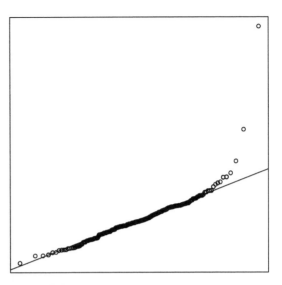

FIGURE 2.15

qqplot of the tiprate from case study i. (axes are omitted as they cannot aid interpretation).

Neither the log-tranformation nor the Box-Cox transformation gives satisfying results in the qqplot. There is one more option left: removal of outliers. Figure 2.16 right shows the qqplot with the three largest values removed — the corresponding boxplot of the variable shows 4 outliers with the smallest of them being only slightly above the upper whisker. The qqplot confirms that this is the best solution of all three approaches. There is no general rule as to how to handle skewed or non-normal data. If normality is a prerequisite of a method a Box-Cox transformation or the removal of outliers might do the job. Both solutions have their problems. Any conclusion drawn from an analysis of a transformed variable must be retranslated into the original domain — which is usually not an easy task. A special handling of outliers, be it a complete removal, or just visual suppression such as hot-selection or shadowing, must have a cogent motivation. At any rate, transformations of data are usually part of a data preprocessing step that might precede a data analysis. Also it can be motivated by initial findings in a data analysis which revealed yet undiscovered problems in the dataset.

Default transformations which standardize data either by mean and standard deviation or onto a $[0, 1]$ range should generally be avoided, as they put all data on scales which can no longer be interpreted.

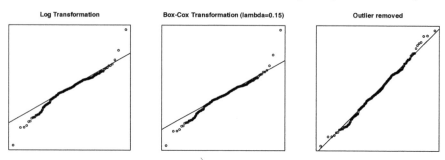

FIGURE 2.16

Three approaches to achieve normality: log-transformation (left), Box-Cox transformation with optimized λ (middle) and a simple exclusion of the three biggest outliers (right).

Categorical Data

At a first sight there might not be much to transform in a categorical variable. But even for categorical data there is a problem similar to skewness. Ordinal, numerical variables such as "number of persons in the family" tend to have the majority of observations distributed among only a few classes.

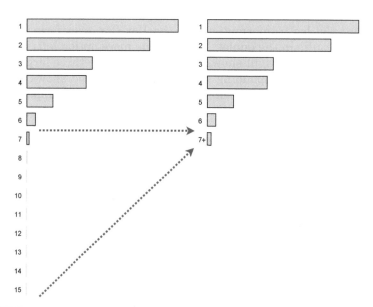

FIGURE 2.17

In the barchart for *Number of Persons in Household*, only 0.5% of all cases make up more than half of the categories.

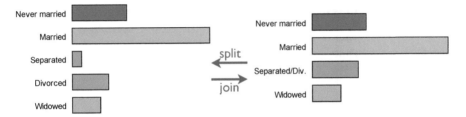

FIGURE 2.18
Joining and splitting classes can be useful even for just a few categories.

Figure 2.17 shows the variable *Number of Persons in Household* for the *Current Population Survey '95* which covers 63,756 of households in the U.S. In this example, only 0.5% of the data make up 8 out of 15 categories. In order to avoid clutter, all categories bigger than 6 can be joined to a new class "7+" as indicated in Figure 2.17 right. One might think that summarizing classes 7 to 15 is a typical preprocessing job. On the other side, it is very efficient to have the ability to decide on such an operation on the fly in an interactive system. Nevertheless, all operations which change data can be a source of errors or at least of misinterpretations and thus should only be applied with much care.

Joining and splitting classes can be very effective even with a few categories. For the same census data used in Figure 2.17 the variable *Marital Status* is shown in a barchart in Figure 2.18. The left barchart shows the two categories *Separated* and *Divorced* as two distinct classes, whereas the right barchart shows the joined version. The two barcharts are not really very different, but all analyses and graphics which are split by, i.e., conditioned on, the 4 or 5 groups of the variable might change substantially.

2.4 Weighted Plots

Basically, one can distinguish three motivations for weighted data. The first is a technical motivation. Whenever we look at purely categorical data, it is not necessary to supply a dataset case by case. A breakdown summary can capture the dataset without loss of any information. Figure 2.19 shows the first 6 lines of the raw data of the *Titanic* dataset, case by case. All six lines are identical as the group of adult male first class passengers who survived has size 140. In this format, the whole dataset has 2,201 entries. The far more efficient version of this information is the summarized data table shown in Figure 2.20. In this representation, the

Class	Age	Gender	Survived
First	Adult	Male	Yes
First	Adult	Male	Yes
First	Adult	Male	Yes
First	Adult	Male	Yes
First	Adult	Male	Yes
First	Adult	Male	Yes
...			

FIGURE 2.19

The first 6 lines of the *Titanic* dataset in raw format, case by case.

dataset has an extra column specifying the size of each group. Since *Class* has four categories, *Age*, *Gender* and *Survived* two each, the dataset will have at most $4 \times 2 \times 2 = 32$ lines. Because 8 of the 32 combinations of the variables do not occur in the data, the dataset can be reduced to 24 lines.

Summarized data tables can be obtained via database queries. A database containing the *Titanic* data case by case can by queried with the simple SQL command.

```
SELECT Class,
       Age,
       Gender,
       Survived,
       count(*)
  FROM Titanic
 GROUP BY Class,
          Age,
          Gender,
          Survived
```

For the *Titanic* dataset, it makes no real difference whether we handle the 2,201 cases in a raw format or the summarized version as long as the software is capable of handling both formats. However, as datasets

Class	Age	Gender	Survived	Count
First	Adult	Female	No	4
First	Adult	Female	Yes	140
First	Adult	Male	No	118
First	Adult	Male	Yes	57
First	Child	Female	Yes	1
First	Child	Female	No	0
...				

FIGURE 2.20

The first 6 lines of the *Titanic* dataset in summarized form.

get really large this difference can become dramatic. The second situation in which weights are introduced is when sampling unequally from a population. Statistics and graphics must then account for the weights. A third reason to use weights is a change of the sampling population. For example a dataset on cancer rates measured on a county level might be weighted with the population of a county in order to switch from the distribution of rates within counties to the distribution of actual cancer cases.

How can statistical graphics incorporate weights? The modification is quite simple for area-based plots displaying counts. Whereas in an unweighted plot bar sizes are proportional to the count in a class, a weighted plot has bar sizes proportional to the sum of the weights in a class. This modification covers plots such as barcharts, histograms and mosaic plots. Glyph-based plots need different modifications. In a scatterplot the point sizes might be adjusted according to the weights. However, this may lead to overplotting and large differences in individual weights could obscure the scatterplot as a whole.

Figure 2.21 gives an example of a weighted histogram compared with the unweighted histogram. The left histogram shows the percentage of votes for G.W. Bush in the 2004 presidential election for the 65 counties in Florida (cf. case study I). The right histogram shows the same plot now weighted by the number of votes in each county. Note that as we change the population from counties to voters — the e.g., rightmost bar ranging from 75 to 80 percent corresponds to 5 counties, this bar represents roughly a quarter million votes (out of almost 7.5 million votes) in the right histogram — the y-scales are no longer comparable. Switching from unweighted to weighted histograms, i.e., from counties to voters, shows that Bush's support was stronger in the less populated counties since high percentages are downweighted and low percentages are upweighted.

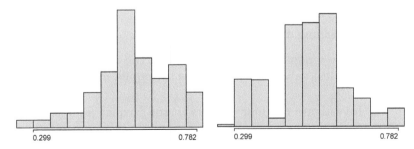

FIGURE 2.21
Histograms of the percentage votes for G.W. Bush in the 2004 presidential election in Florida. Left: unweighted counts by county, Right: percentages weighted by the number of voters.

Exercises

2.1. Barcharts and Spineplots
For the *Tipping* data in case study i

(a) Use barcharts and spineplots to investigate on what days smoking parties are most common.

(b) What problem arises when we look at the size of smoking parties?

2.2. Dotplots, Boxplots and qqplots
For the *Tipping* data in case study i

(a) Compare the benefits of dotplots and qqplots for the variable *Tip in USD*.

(b) What is the percentage of points classified by a boxplot to be an outlier if the underlying distribution is assumed to be standard log-normal?

(c) Can we expect to observe outliers in a boxplot at the steep side of a skewed distribution?

2.3. Histograms & Spinograms
Recreate the graphics from Figure 2.11. What can be said about the rental prices of apartments in buildings built before World War II?

2.4. Density Estimators

(a) Create a histogram with a density estimate for the tiprate from case study i. How do the outliers influence the estimate?

(b) What does the shape of a density estimate look like for the bandwidth $c \to \infty$?

2.5. Transformations
For the *Tipping* data in case study i

(a) Draw a qqplot for the tiprate with the three largest and the smallest values removed. Does the qqplot improve over Figure 2.16?

(b) Investigate the four outliers in the boxplot for *Tiprate*. Why could these cases be treated separately or even neglected?

2.6. Weighted Plots

(a) Create the two graphs in Figure 2.21 of the Florida election data for John F. Kerry. Do these graphs yield a consistent interpretation to what we learned from Figure 2.21?

(b) Create a summarized version of the data from case study B. How many lines has the file?

3

Interactions between Two Variables

This chapter will introduce all important two-dimensional plots. Furthermore, it will show how the univariate plots introduced in the previous chapter are most effectively used to explore relations between two variables by using linking and highlighting. Sometimes there is no single best way to investigate an association, and some plots might be extended to even more dimensions than two.

3.1 Two Categorical Variables

Highlighting and linking of two barcharts can already tell a lot about the relation of two categorical variables. If one of the variables has only two categories, linked highlighting can capture all information of the joint distribution. Figure 3.1 gives an example of this situation. The joint dis-

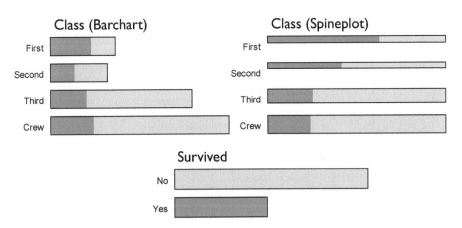

FIGURE 3.1
Linking the survival information of the Titanic dataset (case study D) to the *Class* variable shows the joint distribution of the two variables.

tribution of the variable *Class* and *Survived* from the Titanic case study
has 8 values. The marginal distributions are depicted in barcharts. By
selecting the surviving passengers, all 8 classes can be investigated in the
barchart or spineplot, depending on whether absolute or relative numbers
are of interest.

Note that the selection of a bar c_i in the variable *Class* does not allow us
to investigate the joint distribution, because it only shows the probability
$P(\text{"Class"} = c_i \mid \text{"Survived"} = s_j); i = 1, \ldots, 4; j = 1, 2$ for one class out of
the four.

Mosaic Plots

A spineplot with highlighting can already be regarded as a two-dimensional mosaic plot. Although mosaic plots can show far more than just two
variables, we start with the two-dimensional case here. Mosaic plots are
defined recursively, i.e., each variable that is introduced in a mosaic plot
is plotted conditioned on the groups already established in the plot. As
with barcharts, the area of bars or tiles is proportional to the number of
observations (or the sum of the observation weights of a class). The direction along which bars are divided by a newly introduced variable is usually alternating, starting with the x-direction. A more formal description
of the construction of mosaic plots can be found in Chapter 4. The corresponding mosaic plots for the data from Figure 3.1 are plotted in Figure
3.2. The left mosaic plot shows *Class* first along x and *Survived* second
along y, i.e., the survival information is plotted conditionally on the class
information. This view corresponds to the spineplot in Figure 3.1. Since
mosaic plots are defined recursively, exchanging the two variables will

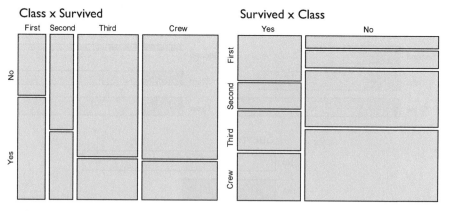

FIGURE 3.2
The two mosaic plots for the variables *Class* and *Survived* from the Titanic case study.

not simply result in a transposed plot, but it will give an alternate view of the data instead. The right mosaic plot in Figure 3.2 has *Survived* first and *Class* second. It displays the distribution of the classes according to the passenger survival.

Mosaic plots are very powerful for the investigation of high-dimensional interactions in categorical data, thus they will be discussed more deeply in Chapter 4. At this point we will focus on variations of mosaic plots, which are especially useful for two-dimensional data with more than just a few categories.

Fluctuation diagrams

Mosaic plots become more difficult to read for variables with more than two or three categories. One way out is to assign a constant space for all possible crossings of categories. This way, the data from the $r \times c$ table are plotted in a table-like layout. Whereas this regular layout makes it much easier to compare values across rows and columns, the plot space is used less efficiently than in a mosaic plot. Fluctuation diagrams share the property of proportional-sized tiles with mosaic plots, but are no longer defined recursively. Instead, the shape, i.e., the aspect ratio, of all tiles is equal. The largest cell fills its slot completely, smaller cells are scaled accordingly. Figure 3.3 shows a fluctuation diagram for *Boat Sequence No.* vs. *Class* for Titanic passengers who entered a life-boat. A clear pattern emerges, depicting how and in which order the different classes have been served. Fluctuation diagrams scale both the width and the height of the tiles, assuring that even small tiles are well visible. A drawback is that a comparison across rows and columns is not directly possible. For instance, a bar half the height of a neighboring bar is only a quarter of the size.

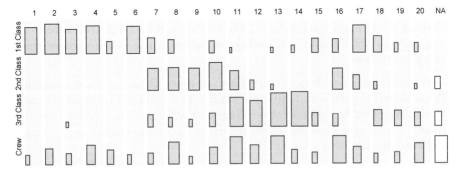

FIGURE 3.3

A fluctuation diagram for *Boat Sequence No.* vs. *Class* for Titanic passengers who entered a life-boat.

Multiple Barcharts

Multiple barcharts scale only the height of the bars. Thus values can be easily compared across rows. The drawback is that small bars will often be reduced to a height of only a single pixel. In Figure 3.4 a multiple

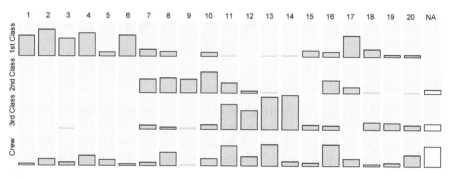

FIGURE 3.4
A multiple barchart view for the same data as shown in Figure 3.3.

barchart is shown for the same data as in Figure 3.3. As the differences between large and small cells are emphasized, the pattern is now even more pronounced.

Alternatively, a special form of zooming — the so-called censored zooming, which is introduced in Chapter 7 — can also be very helpful in gaining insight into small cells of a fluctuation diagram or multiple barchart.

The layout of a fluctuation diagram or multiple barchart must respect the aspect ratio of the tiles to achieve good readability. The so-called "banking to 45°", introduced by Cleveland (1985), can be applied to fluctuation diagrams and multiple barcharts as well. The average aspect ratio of a tile should be one, i.e., the tile should be a square. Often this can only be achieved by reordering variables in the plot and/or resizing the plot window.

Whereas mosaic plots can display far more than just two variables, fluctuation diagrams and multiple barcharts should not incorporate more than two variables. Fluctuation diagrams in higher dimensions are difficult to read and challenging to interpret and thus should only be used with much care.

Highlighting in mosaic plots and its variants is straightforward. The highlighting direction is always orthogonal to the last splitting direction of a mosaic plot, e.g., in the mosaic plots in Figure 3.2 each cell would be highlighted from left to right.

3.2 One Categorical Variable and One Continuous Variable

Although the role of the two categorical variables can often be interchanged, the relationship between a categorical and a continuous variable has usually a defined direction.

Categorical → Continuous

Given a categorical variable, we are interested in the distribution of the continuous variable for different categories of the categorical variable. A boxplot y by x displays the distributions of a continuous variables conditioned on the categories of the categorical variable. Figure 3.5 shows a boxplot of *Rent per m^2* by *No. of Rooms* for the data from case study E, along with a barchart showing the sizes of the groups. Figure 3.5 shows how the rent per area increases for apartments with fewer rooms. Apartments with 6 rooms are relatively the cheapest, but only 0.68% of all cases fall into this category.

With many categories, a boxplot y by x is the most efficient way to

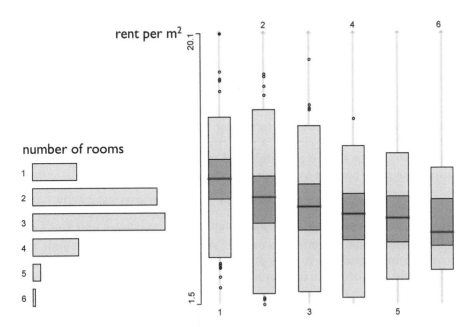

FIGURE 3.5
The distribution of the continuous variable is displayed in boxplots conditioned on each level of the categorical variable.

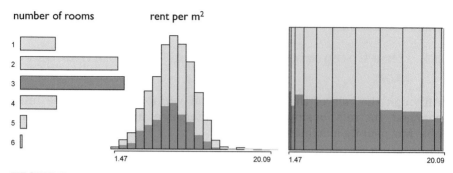

FIGURE 3.6

Investigating the rental price per square meter for 3 room apartments. Histogram and spinogram are set to left closed €1 intervals.

compare the distributions across the different categories. However, box-plots show only selected features of the distributions, namely, the median, the IQR and outliers. Further structural details of the distributions and a comparison of a subgroup against the complete sample cannot be observed from such a plot.

If one restricts the investigation to a single category, a linked barchart and histogram/spinogram can give further insight. Selecting a single category in the barchart will highlight the conditional distribution of that group in the histogram. As shown in Chapter 2, it is difficult in a histogram to compare the highlighted distribution to the distribution of the complete sample. Switching to the spinogram view helps to overcome this problem.

Investigating the example in Figure 3.6 we should formally describe what a selection in the bars represents. Looking at the whole highlighted portion of the histogram gives us the conditional distribution for all three-room apartments. Looking at a particular bar, for instance, the interval [8.00, 9.00), gives us the probability of selecting a three-room apartment given the price per square meter is between €8.00 and €9.00:

$$P(\text{``No. of Rooms''} = 3 \mid \text{``Price per m}^2\text{''} \in [8.00, 9.00)) = 41.03\%$$

Since we want to make a statement about the distribution of the price of three-room apartments, the probability we can read from a single bar is not helpful, because it gives a probability that is conditioned on price. Thus it is only useful to interpret the selected subgroup in the histogram and spinogram as a whole.*

Density estimations can be used in order to reduce the variability of the different bars (changing bar width and starting point can lead to a strong

---

*Figure 3.6 shows the joint distribution of the binary variable indicating the selection and the continuous variable discretized to the 20 classes of the histogram/spinogram.

variation in the shape of the distribution). Figure 3.7 shows the same histogram and spineplot as in Figure 3.6 with a density estimator superposed. A normal kernel has been used and the bandwidth corresponds

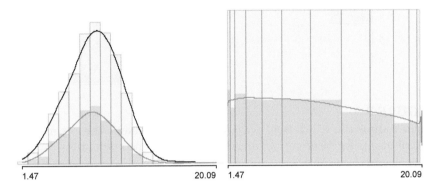

FIGURE 3.7
Adding density estimates to histogram and spinogram delivers a smoother estimate.

to the bin width, i.e., €1. Adding a density estimator in a histogram for the complete sample is straightforward, as long as the y-axis is scaled to relative frequencies, which correspond to probabilities. A problem occurs when the density estimate for the highlighted cases is added. As density estimates are defined to integrate to one, the density estimate will not match the selected points in a histogram, but compare to the density estimate of the complete sample instead. Although this might still be useful, it is not what the user expects. The estimate must be scaled in order to make the density estimate of the selected points match the selected cases in the histogram. For a sample of size n and a group size n_s of the selected cases, the scaling factor f is $\frac{n_s}{n}$. Note the high variance of the density estimator for large values where the estimates are based on only a very few points. Switching to the conditional view of the spinogram needs no further adjustment of the estimated density of the selected subgroup.

So far, only a single group could be investigated in the histogram or spinogram, which is a drawback compared to the boxplot y by x as shown in Figure 3.5. Using color-brushing allows us to color-code the information of several groups in a spinogram simultaneously. Figure 3.8 shows the conditional distributions of all 6 groups of the variable *No. of Rooms*. In this view, we see the distribution of the categorical variable conditioned on a continuous variable.

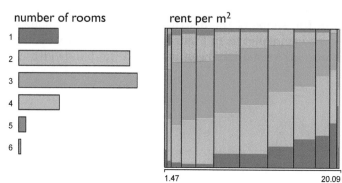

FIGURE 3.8
Color-coding the 6 levels of *Number of Rooms* shows the 6 conditional
distributions for *Rent per m²* as stacked bars in the spinogram.

Continuous → Categorical

Although the distribution of a categorical variable conditioned on a continuous variable is fully displayed in a color-brushed histogram/spinogram
as shown in Figure 3.8, a more flexible setting may be desirable to investigate the influence of a continuous variable on a categorical variable.
The typical setup is a histogram holding the continuous variable and a
spineplot holding the categorical variable. The histogram offers a very
convenient way to select specific intervals of interest while the distribution of the categorical variable can be monitored in the spineplot. Again,
we need the distinction between the interpretation of a single bar and the
highlighting in the spineplot as a whole. The highlighting in a bar of the
spineplot shows the probability that cases fall into the interval selected

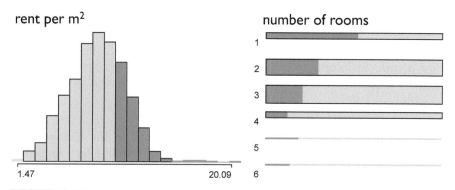

FIGURE 3.9
Selecting intervals in the histogram allows us to monitor that group in
the spineplot.

in the histogram given the category of the bar. In the example in Figure 3.9 all apartments with a *Price per m²* of more than €10 are selected. Looking at the bar of three-room apartments, we read

$$P(|\text{"Price per m}^{2}\text{"} \geq 10 \mid \text{"No. of Rooms"} = 3) = 20.7\%$$

This probability is conditioned on the categorical variable, although we wanted to learn about the distribution of the categorical variable given a certain range in the continuous variable. This information can be read from Figure 3.9 by interpreting all selected proportions in all bars in the spinogram as a whole. From Figure 3.9 we learn that apartments that cost more than the psychological threshold of €10 are most frequently one-room apartments (more than 50%). The proportion declines to less than 20% for apartments with more rooms.

When no specific interval of the continuous variable is targeted, the dynamic technique of brushing can be used. For a fixed brush size, i.e., a fixed interval size, the brush can be moved along the range of the continuous variable while the conditional distribution of the categorical variable can be monitored in the spineplot (cf. Figure 3.10).

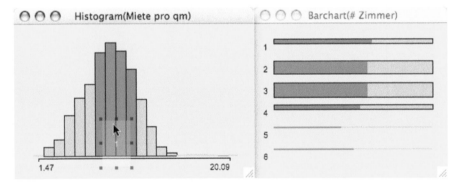

FIGURE 3.10
Using a brush, the investigation from Figure 3.9 may also be performed dynamically for a wider range of intervals.

3.3 Two Continuous Variables

The scatterplot — as the canonical display for two continuous variables — is probably one of the oldest and most familiar plots. In a cartesian coordinate system, a glyph is plotted for all pairs of observations $(x_i, y_i), i = 1, \ldots, n$.

Scatterplots are the ideal tool to investigate the relationship between two continuous variables. Generally, two situations can be distinguished:

- No dependency between the two variables x_1 and x_2 in the scatterplot can be assumed a-priori. In this case, it does not matter which variable is assigned to which axis.

- A functional relationship like $y = f(x)$ is assumed and analyzed. In this case, the dependent variable y is assigned to the vertical axis, and the independent variable x to the horizontal axis.

Many different structures can be investigated in a scatterplot. A general association or pattern may have local anomalies or may be of a different type for different subgroups. Outliers and clusters may be visible in a scatterplot which cannot be detected in a univariate view such as a histogram or dotplot.

The association between two continuous variables is often quantified with coefficients such as the classical correlation coefficient or other statistics based on ranks and concordance of pairs (x_i, y_i) and (x_j, y_j) such as

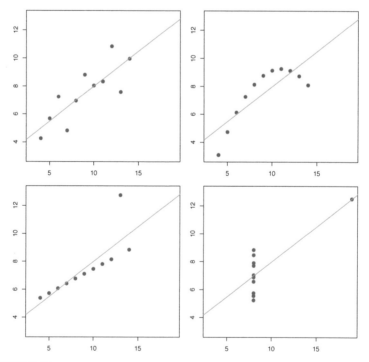

FIGURE 3.11
Anscombe's Quartet, four scatterplots of a very different structure, all having the same linear relationship between x and y.

Spearman's rank correlation. Scatterplots capture the complete structure of the two-dimensional distribution, whereas coefficients can only capture a certain aspect like linear dependence etc.

A famous example — shown in Figure 3.11 — with just 11 points each, is the so-called *Anscombe Quartet*. It shows scatterplots of four datasets, each having the same linear relationship and thus identical correlation r_{xy} and regression coefficients. The upper left plot in Figure 3.11 is the classical situation, where a linear fit seems appropriate. The upper right plot shows a quadratic relationship with no noise, which is not captured correctly by the linear fit. The lower left plot has a perfect linear relationship except for an outlier which biases the fit. The lower right plot does not reveal any relation between x and y, the linear relation is only defined by a single outlier. Although these data are artificial, they all describe general situations which can be quite often found in practice.

Scatterplot Smoothers

In order to study the functional relationship between x and y, it is often helpful to include a scatterplot smoother which tries to estimate a possible functional relationship of the kind $y = \hat{f}(x) + \epsilon$. A simple linear relationship such as a least-square line is often not flexible enough to capture the functional relationship. For the birthweight dataset from case study C, Figure 3.12 shows a least square estimate of a linear function to model the relation between *Gestation* and *Birthweight*. With an R^2 of mere 18% the fit is not very satisfactory as the slope appears to be too small for shorter pregnancies and too large for longer pregnancies. Estimates which fit a polynomial of higher degree are more flexible, but usually not able to capture local structure, either. To achieve a fit that is locally more adaptive, a *loess*-smoother might be applied. A loess-smoother (as defined by Cleveland, 1979) performs a locally weighted regression to estimate a functional relationship. To fit a loess curve at the point x_0, a weighted regression of a polynomial of degree one or two is performed, with the weights

$$T(u) := \begin{cases} \left((1 - |u|)^3\right)^3 & \text{for } |u| < 1 \\ 0 & \text{otherwise} \end{cases} \tag{3.1}$$

as a weight function, such that we get weights at x_0 of the form

$$w(x_0) = T\left(\frac{|x - x_0|}{s}\right) \tag{3.2}$$

where s is the so-called span. This smoothing parameter essentially defines how large is the local region in which neighboring points are given non-zero weight. Locality can be specified in two ways. The above definition uses a fixed interval, which will include a varying number of points

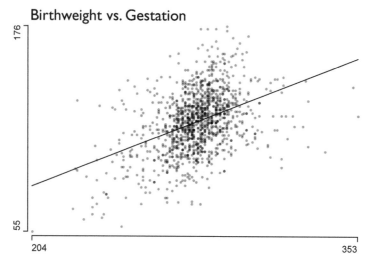

FIGURE 3.12
A simple linear least square estimate to describe the relationship between *Gestation* and *Birthweight* from the birthweight dataset from case study C (note that the two outliers with the smallest gestation times are excluded).

into the estimate for different points x_0. An alternative locality specification is to define a fraction of points which should enter the estimate, leading to a fixed number of observations, but intervals of variable width. Once the weights are defined, the estimate at the points x_0 is

$$\hat{f}(x_0) := X(X'W_0X)^{-1}X'W_0y \qquad (3.3)$$

which is an ordinary weighted regression with weights W_0 at x_0. This means that a weighted regression must be performed for each point to estimate. All estimates are then joined by lines in order to get a sufficiently smooth curve. (Most implementations of loess smoothers offer further robust iterations which down-weight points with large residuals.)

Figure 3.13 shows an example of a loess smother for the data in Figure 3.12. A span of 0.63% of the points has been chosen. This estimate seems to be more sensible than the simple linear fit. The slope is almost constant up to a gestation time of 295 days — i.e., 2 weeks later than the usually expected time of a pregnancy — and remains at an almost constant level for times longer than 295 days.

Choosing the "right" smoothness is a challenging problem, and much literature can be found presenting optimal choices of parameters under different circumstances. Scatterplot smoothers as an exploratory tool should aid the analyst to better judge a possible relationship between the variables x and y. Thus it is desirable to interactively increase and

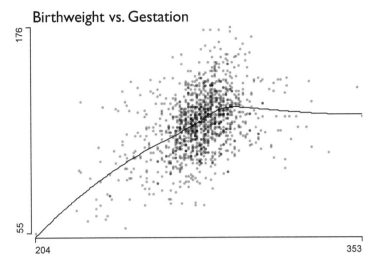

FIGURE 3.13
A loess smoother for the birthweight dataset with a span of 0.63.

decrease the amount of smoothness for a chosen scatterplot smoother. Figure 3.14 shows two examples of more extreme choices of the smooth-

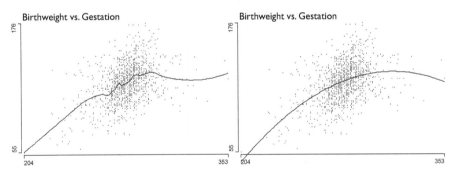

FIGURE 3.14
Two loess smoothers with different smoothing parameters. In the plot on the left, a span of 0.2 undersmooths the data, whereas in the right-hand side plot a parameter of 3.75 oversmooths the data.

ing parameter. The left-hand plot uses a parameter of 0.2 and appears to be undersmoothing within the range of 265 to 300 days. The right-hand plot in Figure 3.14 uses a span of 3.75 (spans bigger than one always use all points and increase the weights further as the span grows) which essentially does not have any locality any more as the weights for all points

x_0 are very close to one. This results in a global estimate of the same kind as the "base"-estimate, which was a cubic fit in this case. The cubic fit is clearly superior to the linear fit, but still too inflexible when compared to the fit in Figure 3.13. It is difficult to give general rules as to how much smoothing is the "right" amount. Being able to modify the smoothing parameter interactively and instantaneously observing the change help very much in understanding the relationship between the two variables in the scatterplot.

The loess smother is an exploratory tool and thus it is infeasible to quantify its variability without computationally expensive methods such as bootstrapping. Smoothing splines provide a parametric smoother which is still locally adaptive in contrast to polynomials. The solution $\hat{f}$ of the minimization problem

$$RSS(f, \lambda) = \sum_{i=1}^{n} \{y_i - f(x_i)\}^2 + \lambda \int \{f''(t)\}^2 dt, \qquad (3.4)$$

leads to a natural cubic smoothing spline. Cubic splines use piecewise polynomials with a constraint of the first two derivatives being continuous. A natural spline adds the further constraint that the function f is linear beyond the bounds of the data. A natural cubic spline can be reformulated to

$$\hat{f} = S_\lambda y \qquad (3.5)$$

where S_λ is a $n \times n$ matrix, i.e., a linear operator. The trace of S_λ, $tr(S_\lambda)$ can be regarded as degrees of freedom of the fitted function and thus allows an equivalent interpretation to other statistical models. For more details on smoothing splines see Section 5.4 in Hastie et al. (2001).

Making a smoother more flexible will also introduce more variability due to the bias-variance-tradeoff. Adding confidence intervals to a scatterplot smoother can help to assess the variability of the estimate. In principle, any smoother can be written in the form of equation 3.5. A pointwise confidence interval then can be estimated via the diagonal elements s_{ii} from S_λ, which are essentially proportional to the width of the confidence interval.

Figure 3.15 shows an example of a smoothing spline with 5 degrees of freedom for *Birthweight* vs. *Gestation*. A 95% pointwise confidence interval has been added to the plot, indicating the variability of the fit. The confidence interval is very small in the center region of the plot. The fewer points are available to estimate $\hat{f}$, the wider the confidence interval gets. The confidence band at both ends of the x-axis of Figure 3.15 widens. On the left side — for very short gestation times — the interval would also cover the linear fit from Figure 3.12. For larger values of the variable *Gestation*, a constant estimate, as we found in Figure 3.13, is within the confidence band. This highlights a common problem of smooth-

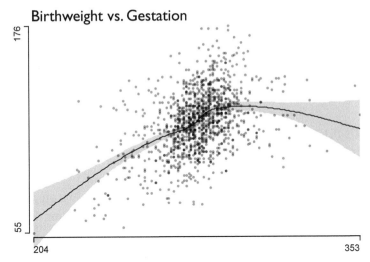

FIGURE 3.15

A natural cubic smoothing spline with 5 degrees of freedom for the birth-weight data from case study C.

ing functions. At the ends of an interval, there are usually only a few values to draw from and additionally the estimate is only based on points to one side of x_0. Thus the result of smoothers at the ends of the estimation interval should always be looked at with great care.

Density Estimation

The problem of overplotting within point-based plots — or more generally in all glyph-based plots — was already dealt with for dotplots in Section 2.2. Although overplotting is less critical in scatterplots, it will occur sooner or later depending on the number of points plotted and the discreteness of the data. Whereas jittering the data points orthogonally to the one dimension in a dotplot did not change the values along the axis of interest, jittering data points in a scatterplot will introduce a "graphical bias" which should be avoided. Furthermore, jittering only helps to spread points where data are accumulated, but cannot indicate regions of high density properly.

A very effective way to handle overplotting is to use α-transparency (often also referred to as α-blending). When using α-transparency, each object has an opacity of $\alpha\%$ (i.e., a transparency of $1 - \alpha\%$). A single point will thus add less "ink" to a plot, but whenever two or more points are added on top of each other, the "ink" will also add and result in a darker area. An example of how α-transparency works for a few objects can be found in Figure 3.16. All circles in Figure 3.16 use a transparency of

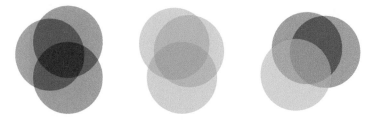

FIGURE 3.16

An example of how objects' colors blend, when α-transparency is applied.

50%. The leftmost group are all black, the middle group is all red, and the right group shows a red circle on top of two black ones. How do the colors blend exactly? The alpha value is the degree of "opacity," and is used to describe how much of the source color is blended with the destination color to create the output color according to

$$col_{\text{dest}} := \alpha \cdot col_{\text{source}} + (1 - \alpha) \cdot col_{\text{dest}} \quad col \in [0,1]. \quad (3.6)$$

For the left example in Figure 3.16 a single point has 50% opacity, adding two points will result in $0.75 := 0.5 \cdot 1 + (1 - 0.5) \cdot 0.5$ which is 75% opacity. Adding a third point on top will give $0.875 := 0.5 \cdot 1 + (1 - 0.5) \cdot 0.75$, i.e., 87.5% opacity. For a single component color — black, red, green or blue — the resulting colors are easy to interpret (cf. Figure 3.16, middle). Mixing colors, as in Figure 3.16 right, is harder to perceive and the resulting colors might be hard to interpret. Using α-transparency is an ideal means to

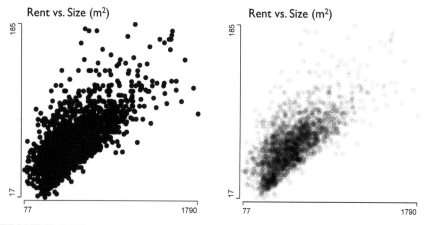

FIGURE 3.17

The same scatterplot of *Rent* vs. *Size* with different values of α-transparency — 100% opacity left; 16% opacity right.

easily distinguish areas of low and high density in scatterplots. Varying point-size and transparency can reveal density structures in a scatterplot. Figure 3.17 shows a scatterplot for *Rent* vs. *Size* for the data from case study E. The left plot uses solid black dots as a glyph, whereas the right plot uses black dots with an α-value of 16%, i.e., 84% transparency.

Depending on the number of observations in a scatterplot and the density structure as such, the values of point-size and α-value have to be selected, such that the density structure can be observed most effectively. Figure 3.18 shows a matrix of three different point sizes — 3, 5 and 7 (from left to right) — and three different α-levels — 8%, 50% and 92% (from bottom to top) for the scatterplot of the variables *oleic* vs. *linoleic* from case study H. The two most extreme choices of point-size and α-value (point-size: 3, α-value: 8% and point-size: 7, α-value: 92%) can hardly show any density information. Most other settings in between show several clusters corresponding to the different groups in the dataset.

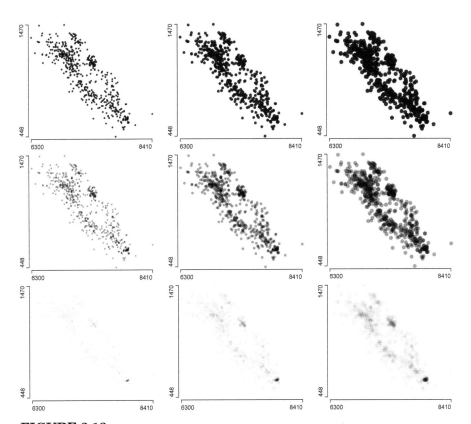

FIGURE 3.18
A matrix of three different point sizes — 3, 5 and 7 (from left to right) — and three different α-levels — 8%, 50% and 92% (from bottom to top).

A special feature in this scatterplot is the strong accumulation of cases at one point in the lower right. More than 6% of the data fall onto nearly the exact same location in the scatterplot, making up almost 2/3 of all *Umbrian* oils. This feature is only visible when using a relatively small α-value, as shown in the bottom row of Figure 3.18.

A further discussion of α-transparency and binning methods for scatterplots can be found in Chapter 9 dealing with large datasets.

Exercises

3.1. 2-Way mosaic plots
   Create all 2-way mosaic plots for the Titanic dataset.

   (a) How many different 2-way mosaic plots can be generated for this dataset?

   (b) Describe what you can read from the most important plots.

   (c) Which plots would you choose to describe the data sufficiently and explain why.

3.2. Simpson's Paradox
   Simpson's Paradox can be characterized by:

   > *Simpsons Paradox refers to the reversal of the direction of an association when data from several subgroups are combined to a single group, which then has the opposite association than all the subgroups.*

   (a) Search the internet for an example of a three-way table, showing Simpson's Paradox, and create the appropriate datafile.

   (b) Create the appropriate mosaic plots, to observe the overall effect (2 variables) and the effects within the groups (3 variables).

   (c) Describe how Simpson's paradox is manifested in a mosaic plot.

3.3. Boxplots y by x
   For the tips data from case study i,

   (a) Create a boxplot *Tip in USD* by *Size of Party*, a barchart for *Size of Party* and a histogram for *Tip in USD*.

   (b) Select different groups of party sizes and observe the highlighting in the histogram.

   (c) Compare advantages and disadvantages of a boxplot vs. an approach via selection and highlighting.

3.4. Spineplot and Spinogram
For the tips data from case study i,

(a) Create a barchart for *Size of Party* and a histogram for *Tip in USD*.

(b) Select and brush different intervals within the histogram and observe the highlighting in the barchart and spineplot.

(c) Color brush according to the barchart. Describe the advantages and disadvantages of selection and color brushing for this application.

3.5. Scatterplot smoother
Describe formally a natural cubic spline with one degree of freedom. How does it compare to the least square estimate of a linear fit?

3.6. 2-dimensional density
For the tips data from case study i,

(a) Create a scatterplot of *Tip in USD* vs. *Bill in USD*.

(b) Use different settings of point-size and α-value in the scatterplot to explore the rounding of the tip to full, half and quarter dollar amounts. Can you see more than in the corresponding histogram for *Tip in USD*?

(c) Do other two-dimensional structures become visible at certain settings?

4

Multidimensional Plots

So far, none of the introduced plots displayed more than two variables at a time. The interaction techniques introduced in Chapter 1 were applied to the plots shown in Chapter 2 and Chapter 3. Selection and linking increased the dimensionality to two, three or even four dimensions. This chapter introduces plots for displaying high dimensional data and discusses their benefits when used in an interactive environment.

4.1 Mosaic Plots

Chapter 3 introduced mosaic plots and their variations to display two-dimensional categorical data. The recursive definition of mosaic plots permits us to display even more dimensions.

Construction

Figure 4.1 gives an example of the construction of a four-dimensional mosaic plot for the Titanic passenger data from case study D. Starting with a rectangular area, the area is divided into proportionally sized bars along the *x*-axis for all categories of the first variable (*Class* in this case). For better visual perception, a small gap is introduced between the bars. Note that this view — shown in Figure 4.1 upper left — is identical to a spineplot.

  In order to introduce a second variable, the bars for all categories of the first variable are subdivided according to the category counts for the second variable, conditioned upon the category of the first variable. The splitting direction for the second variable is along the *y*-axis. For the Titanic example this means that the bars for the different classes are subdivided according to the distribution of *Age* within each category of *Class*. The resulting two-dimensional mosaic plot is shown in Figure 4.1 upper right. Note that there is an empty bin for children in the crew.

  Adding a third variable follows the same recursive scheme as described above. The splitting direction swaps back to the *x*-axis. In the example of Figure 4.1 all intersections of *Class* x *Age* are divided along the *x*-axis

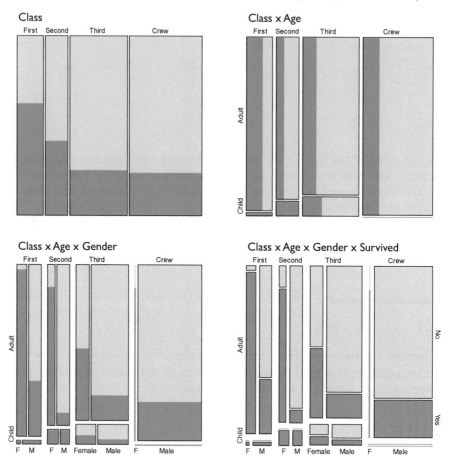

FIGURE 4.1

Step by step construction of a mosaic plot for the Titanic dataset (case study D). All surviving passengers are highlighted in all plots.

according to the distribution of *Gender* within this intersection. This plot, shown in Figure 4.1 lower left, shows that there are only very few female crew members.

The next, and last, variable is then split along the y-axis again. The fourth variable in the Titanic example is *Survived*. It splits all crossings of *Class* × *Age* × *Gender* according to whether the passengers survived or not. The final plot, with all four variables included, can be found in Figure 4.1 lower right. This last plot also reveals the information highlighted in all four plots: surviving passengers. Highlighting in mosaic plots is always plotted orthogonal to the last splitting direction, i.e., if the last splitting direction was along x highlighting is done along y, and vice versa. Looking at this plot, we find an empty red rectangle for non-adult crew

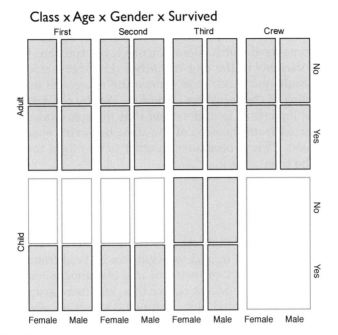

FIGURE 4.2
The four-dimensional mosaic plot for the Titanic dataset (case study D)
in same bin-size view.

members. Since this intersection is empty it will remain empty even as
further variables are added. We allocate for it the same amount of space
that is used by the gaps between the classes of non-empty intersections.

Figure 4.2 is a mosaic plot in the 'same bin-size' view and shows the
same variables as the mosaic plot in Figure 4.1 lower right. In a same
bin-size view all tiles are of the same size and the information is reduced
to the binary information of whether a cell is empty or not. Due to the
"Curse of Dimensionality" — which essentially states that the space gets
more sparse as the number of displayed dimensions increases — the number of empty cells will increase as more variables are added to a mosaic
plot. In this case, it is of interest to investigate the structure of empty
cells. In principle, two kinds of empty cells can be distinguished: **structural** zeroes, i.e., it is technically impossible to observe cases in a certain combination of categories, and **random** zeroes, i.e., an intersection
is empty due to a very small probability for that particular group. From
Figure 4.2 we learn that all children in the first and second class survived
and that there are no children in the crew at all.

Looking back at Figure 4.2, we see the size of the gaps between categories of the first variable increase as more variables are included in the
plot. The underlying principle for the width of the gaps between cate-

gories is as follows: the gap between the categories of the last variable in a plot, i.e., the last variable in the recursion, is the smallest, and has always the same fixed width. Going up one level, the gaps become wider. For the first variable in the plot it reaches the largest size. This makes reading a mosaic plot much easier since the categories are grouped together according to the recursive structure. For the construction of a mosaic plot, it is important to understand that the space taken by the gaps is not subtracted from the area of the tiles, but extra space is added to the plot instead. This is necessary in order to maintain the area proportionality of the tiles.

Ordering

Due to their recursive definition, switching the order of variables in a mosaic plot has a strong impact on what can be read from the plot. For instance, exchanging the two variables in a two-dimensional mosaic plot results in a completely new plot rather than in a mere graphically transposed version of the original plot. Figure 4.3 shows an alternative order

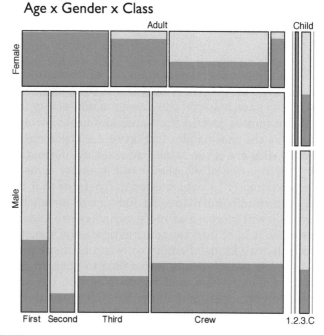

FIGURE 4.3

The same mosaic plot as in Figure 4.1 (lower left), now with a different order, focuses on different features in the dataset.

of the variables of the mosaic plot in Figure 4.1. *Age* and *Gender* are put first, such that the association between *Class* and *Survived* can be compared for each of the four combinations of the variables. The information for children — which represents less than 5% of the data — seems to diminish in this view and is less relevant for the interpretation of the plot. For a deeper interpretation of the dataset and the related mosaic plots please refer to case study D.

If no specific question determines the order of the variables in a mosaic plot several recommendations can be used to construct a classical high-dimensional mosaic plot (see also Theus, 2007):

- The association of the first two or the last two variables in a mosaic plot can be investigated most efficiently. Thus, an interaction of interest should be put into the last two positions of the plot. Variables which condition an effect should be the first in the plot. Figure 4.1 shows the interaction between *Class* and *Age* as these variables enter the plot first. For each of the combinations, the association between *Gender* and *Survived* can be compared since these are the last two variables.

- Put variables with only a few categories first in order to avoid unnecessary clutter in a mosaic plot of equally important variables.

- If combinations of cells are empty, seek variables which create empty cells at high levels in the plot in order to reduce the number of cells to be plotted. Empty cells at a higher level are not divided any further, thus gathering many potential cells into one. In Figure 4.1 the empty cell of non-adult crew members is not split any further, thus gathering four cells into one.

- If the last variable in the plot is a binary factor one can reduce the number of cells by linking the last variable via highlighting. This is the usual way of handling categorical response models.
  The two lower graphics in Figure 4.1 show essentially the same information, because *Survived* is a binary factor and thus can be sufficiently displayed via highlighting.

- Subsets of variables may reveal features far more clearly than all variables at once. In an interactive mosaic plot it is possible to add, drop or change variables displayed in a plot. This is very efficient when looking for potential interactions between variables. The stepwise development of the four-dimensional mosaic plot in Figure 4.1 is far easier to follow than directly looking at the plot holding all four variables.

Although these guidelines turn out to be helpful, many different views may be visited in an actual data analysis in order to find the best representation for a specific feature.

Mosaic Plots and Categorical Models

Mosaic plots are not limited to displaying counts of categorical variables. Any positive variable can be used as weight for the different categories in a mosaic plot (cf. also Section 2.4). The most interesting application is to substitute counts by estimates of a statistical model for categorical variables. For instance, for a mosaic plot with l cells displaying n observations, the same bin-size view corresponds to the trivial model

$$\hat{o}_j = e_j = \frac{n}{l}.$$

The above model is usually not very sensible. Thus the model of total independence is usually referred to as the "null-model" for multivariate categorical data. It is defined by

$$\hat{o}_{i_1,\dots,i_k} = e_{i_1,\dots,i_k} = n \cdot \pi_{1,i_1} \cdot \ldots \cdot \pi_{k,i_k},$$

with π_{j,i_j} being the probability of the marginal distribution of the j-th variable and the i_j-th level, estimated by $\dfrac{n_{j,i_j}}{n}$, where n_{j,i_j} is the marginal sum for variable j and level i_j. Figure 4.4 compares the observed data for the first three variables of the Titanic data (left) with the corresponding plot for the model of complete independence (right). Obviously this simple model cannot capture the structure of the data very well. At the very least the empty cell would need to be treated specially.

More complex log-linear models, and their interplay with mosaic plots are discussed in Section 5.3.

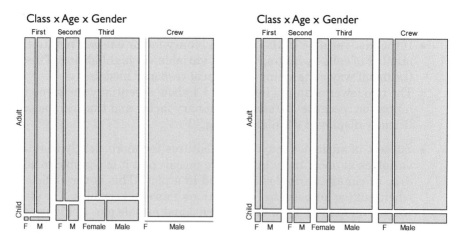

FIGURE 4.4

Left, the same mosaic plot as in Figure 4.1 (lower left). Right, the plot for the same variables, displaying the model which assumes complete independence.

4.2 Parallel Coordinate Plots

Whereas mosaic plots are used to display high-dimensional categorical data, parallel coordinate plots are used for high-dimensional continuous data.

Construction

A parallel coordinate plot draws an axis for each variable in the plot. As the name suggests, all axes are plotted in parallel. For a dataset $X_{i,j}; i = 1,\ldots,n; j = 1,\ldots,p$ with n observations and p dimensions, each observation results in a polyline. The edges of the polygon are the points $x_{i,j}$, which are plotted at axis j and its coordinate value $x_{i,j}$.

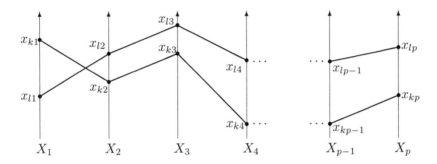

FIGURE 4.5
Two p-dimensional observations $x_k = (x_{k1}, x_{k2}, x_{k3}, \ldots, x_{kp-1}, x_{kp})'$ and $x_l = (x_{l1}, x_{l2}, x_{l3}, \ldots, x_{lp-1}, x_{lp})'$ plotted in parallel coordinates.

Figure 4.5 shows an exemplary parallel coordinate plot for two points x_k and x_l in p dimensions. The coordinate value for each point x_{kj} (and x_{lj}) is plotted on each axis $j, j = 1,\ldots,p$ and then joined by a polyline. From this definition it is obvious that categorical data cannot be displayed in parallel coordinate plots properly since category labels cannot be plotted in a meaningful way along a continuous axis.

Parallel coordinate plots have many interesting geometric properties (for instance, the point-line-duality, cf. Exercise 4.3) and thus were first explored in mathematics (see Inselberg, 1985). The application of parallel coordinates in data analysis was first illustrated by Wegman (1990).

Parallel coordinate plots are often overrated concerning their ability to depict multivariate features. Scatterplots are clearly superior in investigating the relationship between two continuous variables and multivariate outliers do not necessarily stick out in a parallel coordinate plot. Nonetheless, parallel coordinate plots can help to find and understand features such as groups/clusters, outliers and multivariate structures in their multivariate context. The key feature is the ability to select and highlight individual cases or groups in the data, and compare them to other groups or the rest of the data.

Three major aspects of parallel coordinate plots can be distinguished:

- **Multivariate Overview**
  No other statistical graphic can hold so much information at a time than the parallel coordinate plot. Thus this plot is ideal to get an initial overview of a dataset, or at the very least a large subgroup of the variables. Figure 4.6 shows a parallel coordinate plot for the olive data from case study H. Moderate α-blending has been applied (cf. Chapter 9) in order to reduce overplotting and to increase the contrast of high density areas to low density areas. Several features are immediately visible, such as discrete structures for small values of *linolenic* and *arachidic*, the groups apparent at the crossing of the lines between *oleic* and *linoleic* or outliers for *oleic* and *palmitic*.

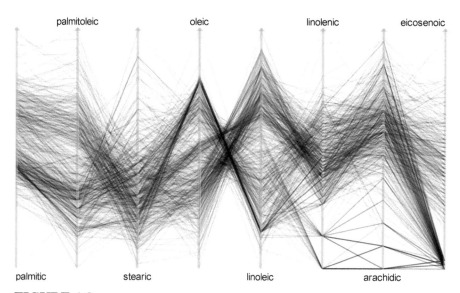

FIGURE 4.6
A parallel coordinate plot for the 8 fatty acids from the olive oil dataset from case study H.

- **Profiles**

  Single items or coherent groups may be selected in a parallel coordinate plot in order to investigate such items or groups for all variables in the plot. In Figure 4.6 we noted the outlier for *oleic*. Highlighting this outlier reveals that it is also an outlier for *palmitic*. Figure 4.6 also showed at least three groups apparent as crossings between the variables *oleic* and *linoleic*. Selecting one of the groups highlights a very consistent and narrow band for most of the 8 variables in the plot (as shown in Figure 4.7).

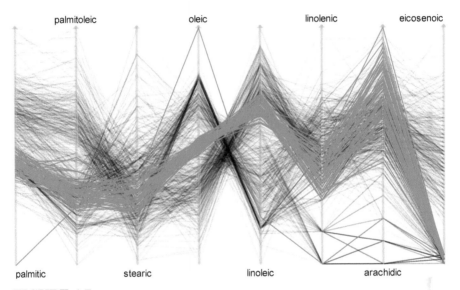

FIGURE 4.7
The same parallel coordinate plot as in Figure 4.6 with a single outlier and a whole group highlighted, profiling their values across all variables.

- **Monitor**

  Multivariate techniques often summarize or classify many variables to only a few groups or factors (e.g., cluster analysis or multi-dimensional scaling). Parallel coordinate plots can help to investigate the influence of a single variable or a group of variables on the result of a multivariate procedure. Plotting the input variables in a parallel coordinate plot and selecting the features of interest of the multivariate procedure will show the influence of different input variables. Figure 4.8 shows the result of a multi-dimensional scaling (MDS) of the 8 fatty acids of the olive oil data in two dimensions. A central cluster in the MDS plot has been selected. In the parallel coordinate

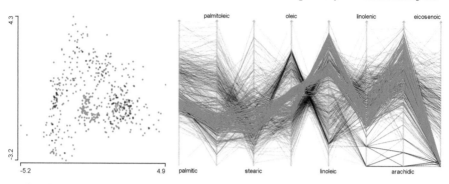

FIGURE 4.8
The first two axes of a 8-dimensional MDS in a scatterplot, linked to the 8 variables in a parallel coordinate plot.

plot, this cluster turns out to correspond to two of the groups which could be identified in Figure 4.6.

Orderings

One big advantage of parallel coordinate plots over scatterplot matrices (i.e., the matrix of scatterplots of all variable pairs) is that parallel coordinate plots need less space to plot the same amount of data. On the other hand, parallel coordinate plots with p variables show only $p-1$ adjacencies. However, adjacent variables reveal most of the information in a parallel coordinate plot. Reordering variables in a parallel coordinate plot is therefore essential. Fortunately it does not take $p!$ plots to view all $\dfrac{p(p-1)}{2}$ potential adjacencies in the plot, but only $\lfloor \dfrac{p+1}{2} \rfloor$ permutations (see Wegman, 1990). This is a relatively small number of plots and a simple task when an interactive system creates the necessary permutations, saving the user from entering the different permutations manually via command-line.

Automatic ordering of variables in parallel coordinate plots can be done by various criteria, such as sorting according to the mean, median, standard deviation, inter-quartile range, minimum or maximum, etc. Figure 4.9 was sorted such that variables with the highest correlations are adjacent. Since the sign of the correlation is important for the visual representation of parallel coordinate plots (cf. Exercise 4.3), the axis of the variable *oleic* has been inverted to make all correlations positive. Inverting an axis in a parallel coordinate plot means flipping this axis around its center, rather than a change of the sign of the variable. The plot in Figure 4.9 "untangles" the parallel coordinate plot as much as possible. Although this makes it easier to spot single odd cases in the plot, it is

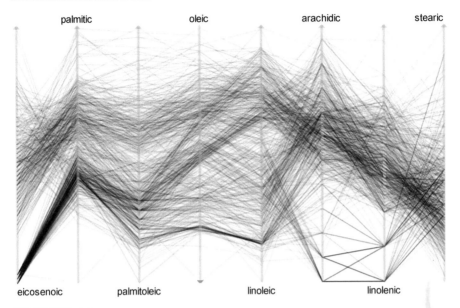

FIGURE 4.9
The parallel coordinate plot from Figure 4.6 with maximized correlations between axes. Note that *Oleic* has been inverted to make all correlations positive.

usually easier to spot groups when the correlation between adjacent axes is negative. Despite the variables *oleic* and *linoleic* being adjacent in both Figures 4.6 (which uses the default order of the variables as they appear in the dataset) and 4.9, the groups are easier to see in Figure 4.6 due to negative correlation.

Scalings and Alignments

So far, we have only looked at parallel coordinate plots which used a min-max-scale, i.e., each axis was individually scaled between its minimum and maximum. If the variables share the same scale (e.g., measures of the same kind at different times or places), it is important to have the option to use the same scale for all axes in one plot in order to compare values across the axes. Figure 4.10 shows the 21 stage times for the 155 cyclists who finished the 2005 Tour de France cycling race in a parallel coordinate plot with individual scales (cf. case study F). (Keep in mind that smaller values correspond to better ranks.) Depending on the kind of stage (flat, mountain or time trial) the data are very much clustered, have extreme outliers or are relatively equally spread. A common scaling of all 21 stages results in Figure 4.11. The time differences within a single

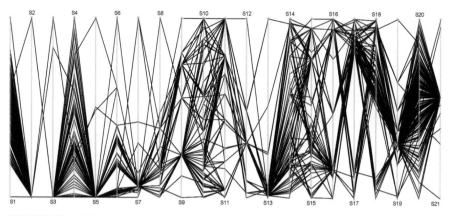

FIGURE 4.10
A parallel coordinate plot for the 21 stages of the 2005 Tour de France
with individually scaled axes (see case study F).

stage are no longer visible for most of the stages due to the large absolute
time differences between a short time trial flat land and a long mountain
stage.

It takes essentially two parameters to scale parallel coordinates: the
offset o_j and the scale s_j such that $y'_{ij} = s_j \cdot y_{ij} + o_j$, $i = 1, \ldots, n$, $j = 1, \ldots, p$. For an individually scaled parallel coordinate plot, distinct s_j
and o_j are used for each variable to completely span the plotting region.
In a parallel coordinate plot with a common scale all $s_j \equiv s$ and $o_j \equiv o$.
In many situations, it is desirable to use individual o_j but a fixed $s \equiv s_j$.
In such a case, the o_j are chosen such that each variable is aligned at

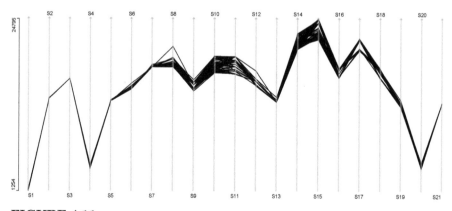

FIGURE 4.11
The same data as in Figure 4.10. This parallel coordinate plot is scaled
to have the same scales on all 21 axes.

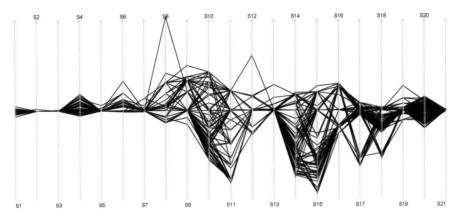

FIGURE 4.12
A parallel coordinate plot for the Tour de France data from case study F. All axes have the same scaling parameter, but are individually shifted by their medians.

an individual offset such as its mean, median, a specific case or simply a constant. Figure 4.12 shows the Tour de France data with common scale but individual offsets, such that each axis is aligned at its median. The median of the stage times usually corresponds to the peloton of this stage and thus has a natural interpretation.

4.3 Trellis Displays

Trellis displays (or Lattice Graphics as they are called in the R software) are an easy to interpret plotting technique to present data in up to seven dimensions. Although there is not much that can be controlled interactively in trellis displays, they are often used to present the results of a data analysis — or parts of it — and thus should not be missing in a book on graphical data analysis techniques.

Construction

Trellis displays were initally introduced by Becker et al. (1996) as a means to visualize multivariate data (see also Theus, 1999). Trellis displays use a lattice-like arrangement to place plots onto so-called panels. Each plot in a trellis display is conditioned upon at least one other variable. The same scales are used in all the panel plots in order to make them comparable across rows and columns.

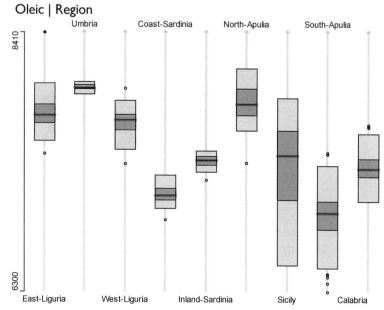

FIGURE 4.13

A boxplot y by x is a simple trellis display.

Probably the simplest example of a trellis display is a box-plot y by x. Figure 4.13 shows a boxplot y by x for the olive oil dataset from case study H. The variable *oleic* is plotted in a boxplot conditioned for all levels of the variable *region*. Because all boxplots share the same scale, values can be easily compared across columns in the plot. This principle can be extended to rows, columns and pages.

By its definition, a single trellis display can hold up to 7 variables at a time. Naturally 5 out of the 7 variables need to be categorical, 2 can be continuous. At the core of a trellis display we find the *panel plot*. The variables plotted in the panel plot (up to two) are called *axis variables* (the current *Lattice Graphics* implementation in R offers higher dimensional plots such as parallel coordinate plots as panel plots, which should be seen as a means to construct small multiples). In principle any arbitrary statistical graphic can be used as the panel plot, but in most cases is no more complex than a scatterplot. All panel plots share the same scale. Up to three categorical variables can be used as *conditioning variables* to form rows, columns and pages of the trellis display. To annotate the conditioning categories of each panel plot, the so-called *strip labels* are plotted atop each panel plot listing the corresponding category names. The two remaining variables — the so-called *adjunct variables* — can be coded using different glyphs and colors (if the panel plot in the trellis

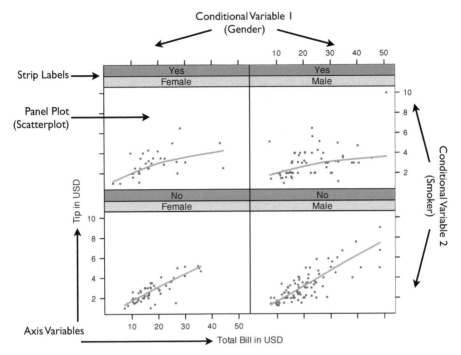

FIGURE 4.14

A trellis display for the tips dataset for the variables *Tip* and *Total Bill*. The conditioning variables have been chosen as *Gender* and *Smoker*. A scatterplot smoother has been added to support the interpretation of the relationship. All elements of a trellis display are annotated.

display is a glyph-based plot).

Trellis displays introduce the concept of *shingling*. Shingling is the process of dividing a continuous variable into — possibly overlapping — intervals in order to convert a continuous variable into a discrete variable. Shingling is quite different from conditioning on categorical variables. Overlapping shingles/intervals lead to multiple representation of data within a trellis display, which is not the case for categorical variables. Furthermore, it is challenging to judge which intervals/cases have been chosen to build a shingle. Trellis displays represent the shingle interval visually by an interval of the strip label. Although no plotting space is wasted, the information on the intervals is difficult to read from the strip label. Despite these drawbacks, there is a valid motivation for shingling, which is illustrated later on. An example of a trellis display which uses shingles can be found in Figure 4.16. The variable *Year of Construction* has been divided into 4 disjoint intervals.

In Figure 4.13 we find one conditioning variable (*Area*) and one axis variable (*Oleic*). The panel plot is a boxplot. Strip labels have been omitted as the categories can be annotated in the traditional way.

An example of a more complex trellis display can be found in Figure 4.14. The scatterplot of *Tip in USD* vs. *Total Bill in USD* is plotted for the tips dataset from case study i. Thus the panel plot is a scatterplot. The axis variables are *Tip in USD* and *Total Bill in USD*. The grid is set up by the two conditioning variables *Gender* along x and *Smoker* along y. There is no adjunct variable included in the plot. The upper strip label shows the category of *Smoker*, the lower strip label shows the category of *Gender*.

In Figure 4.14 the factor *Smoker* seems to have a notable effect on the estimate of the functional relationship between tip and bill-size, whereas the factor *Gender* does not seem to change the estimate.

Models

Trellis displays are an ideal tool to compare models for different subsets. In Figure 4.14 a loess smoother has been superposed onto the scatterplot. This loess smoother is fitted for each panel separately, and thus allows us to judge how well a model fits for different subgroups. By comparing the different fits, the analyst can then decide which factors should be included in a model and which groups or cases might be treated as outliers.

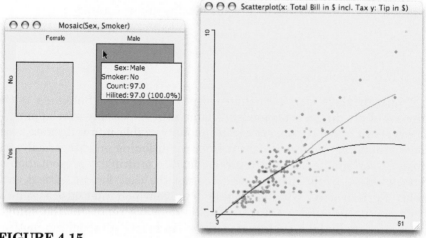

FIGURE 4.15

Selecting male non-smokers in the mosaic plot highlights this group in the scatterplot.

Trellis Displays vs. Interactivity

The conditional framework in a trellis display can be regarded as systematic static snapshots of selection states in interactive statistical graphics.

A single view in a panel of a trellis display can also be thought of as the highlighted part of the graphics of the panel plot for the conditioned subgroup. This can be best illustrated by looking at the tips dataset. Figure 4.15 shows a screen shot of an interactive session. Selecting male non-smokers in the mosaic plot highlights this group in the scatterplot of *Tip in USD* vs. *Total Bill in USD*, which is the same view as in Figure 4.14 lower right. Note that unselected points are shadowed using a small α-value. The view in Figure 4.15 permits us to compare the selected group and its smoother to all data, which is not possible in a trellis display. On the other hand, a comparison to a different conditioning group is more challenging in the interactive set-up, and can only be done by switching back and forth beween the two groups in question.

Selecting a specific subgroup in a barchart or mosaic plot is one interaction. Another interaction would be brushing. Brushing a plot means to move a brush — i.e., an indicator for the selection region — steadily along one or two axes of a plot. The interval selected by the brush can be seen as an interval of a shingle variable (for brushing see also Section 1.2, page 17). When a continuous variable is subdivided into say 4 intervals, this corresponds to 4 snapshots of the continuous brushing process

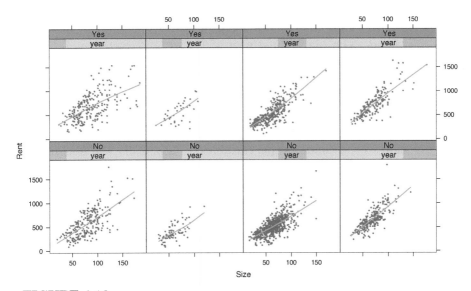

FIGURE 4.16

A trellis display for the *Rent* data from case study E using a shingle. The variable *Year of Construction* has been divided into 4 intervals.

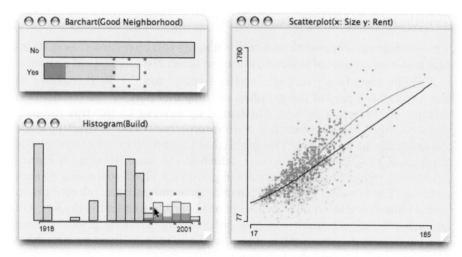

FIGURE 4.17
Selecting the appropriate subset within the conditioning variables yields
the same scatterplot as in Figure 4.16 upper right.

from the minimum to the maximum of that variable. For the same scatterplot used as panel plot in Figure 4.16, Figure 4.17 shows a snapshot
of a brush selecting the largest values of the conditioning variable *Year
of Construction*; the category *Yes* has been selected for the variable *Good
Neighborhood*. The highlighted part of Figure 4.17 corresponds to the
upper rightmost plot in Figure 4.16.

The motivation of shingle variables now is getting more obvious since
they relate directly to this interactive technique. Brushing with linked
highlighting is certainly far more flexible than the static view in a trellis
display. Its biggest advantage over trellis displays is the fact that the
distribution of the conditioning variables — especially the shingles —
is transparent. The quite ragged distribution of *Year of Construction* is
hidden in the trellis display in Figure 4.16, but is immediately visible in
the histogram of Figure 4.17, which is used for brushing.

On the other hand, the trellis display can easily be reproduced in printed
form, which is impossible for the interactive process of brushing.

Exercises

4.1. Zeroes in mosaic plots

(a) Discuss the nature of the zero counts in the mosaic plot for the Titanic data.

(b) Think of examples of structural and random zeroes for other categorical datasets.

(c) Can you think of situations where a strict distinction is no longer possible?

4.2. Double-Decker Plot
Hofmann (2001) defined double-decker plots as mosaic plots which only split along the x-axis.
Create a double decker plot for the Titanic data for the variables *Class, Age* and *Gender* and highlight the survivors.

(a) What can be learned from a double-decker plot which is more difficult to read from a classical mosaic plot?

(b) How does the order of the variables influence the double-decker plot?

4.3. Point-line-duality of parallel coordinates
Generate a sample of 100 points from two variables which follow a strict linear relationship of the form $x_2 = ax_1 + b$. Choose arbitrary values for a and b and plot the resulting values in parallel coordinates.

(a) How does the value of a influence the shape of the parallel coordinate plot?

(b) Add different amounts of noise $\epsilon \sim N(0, c)$ to the linear function. How does this change what you can see in parallel coordinates?

4.4. Cluster in parallel coordinate plots
Generate two two-dimensional samples (X, Y) of size 100 each from $N((\mu_{x1}, \mu_{y1}); I_2)$ and $N((\mu_{x2}; \mu_{y2}), I_2)$. Plot the samples in parallel coordinates. Use $(\mu_{x1}, \mu_{y1}) = (1, 1)$ and

(a) $(\mu_{x2}, \mu_{y2}) = (2, 2)$

(b) $(\mu_{x2}, \mu_{y2}) = (4, 1)$

(c) $(\mu_{x2}, \mu_{y2}) = (4, 4)$

How well are the two groups separable in parallel coordinates and scatterplots? How well are the groups visible in univariate plots?

4.5. Trellis displays

Discuss the following points.

   (a) What is the problem when adding a second conditioning variable to Figure 4.13?

   (b) Why are barcharts not very well suited as panel plots?

   (c) Shingle intervals may be overlapping or non-overlapping, of equal length or of equal case-count. What are the advantages and drawbacks of each approach?

4.6. Shingling

Create the set-up in Figure 4.17 and experiment with different widths of the brush and different intervals.

   (a) Compare the results with the panel plots in Figure 4.16. Which intervals have been chosen in Figure 4.16? Would you propose other intervals, and if so, why?

   (b) Explain the difference between brushing a histogram and a dotplot.

   (c) Can the set-up from Figure 4.17 be reproduced without the use of selection sequences?

5

Plot Ensembles and Statistical Models

Chapter 3 introduced plots and combinations of plots to investigate two-dimensional associations. Chapter 4 presented high-dimensional plots, which can incorporate many variables of the same type.

This chapter will illustrate typical scenarios of combinations of plots — so-called plot ensembles — which can be used to investigate high-dimensional data of certain types most effectively, relying on linking and one-, two- and multidimensional plots. There are certainly far more linked plot combinations than can be presented in this chapter; therefore we focus on typical problems which are often dealt with in classical statistics.

5.1   Response Models

It is often desirable to investigate or model the effect of a set of several influencing factors or variables (inputs) on a single dependent response variable (output)i. Different combinations of plots have proven to be most effective depending on the kind of input and output variables.

A categorical response variable can be displayed using a barchart or spineplot; a continuous response variable is better displayed in a histogram or boxplot. However, the choice of the appropriate graphical representation of the input variables is more important.

Categorical Inputs

When all inputs in a response model are categorical, a categorical plot, i.e., a mosaic plot or a variant thereof is most suitable. Figure 5.1 shows the setting for the Titanic data: the outcome (survived) is selected in a barchart of the variable *Survived*. All influencing factors — *Class, Age* and *Gender* — are displayed in a mosaic plot (cf Figure 4.1 lower left). Each intersection of the three factors is shown with the corresponding highlighting – the proportion of survivors in that group. An advantage of using a mosaic plot is that not only the interaction structure between the influencing factors can be investigated, but also the interaction between the response and the different influencing factors. In Figure 5.1

Class × Age × Gender

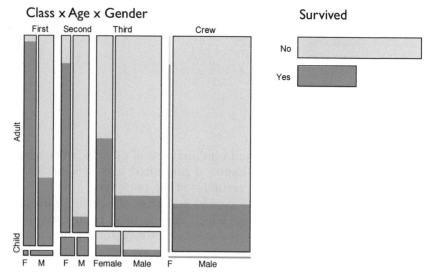

Survived

FIGURE 5.1

Selecting the outcome (survived) in the barchart, while observing the highlighting in combinations of the inputs in a mosaic plot.

we immediately see that survival proportions are higher for females and for "upper" classes.

The disadvantage of this view is that highlighting proportions cannot be readily compared since the absolute sizes of the tiles vary along the highlighting direction. A double-decker plot can be used to overcome this problem. Figure 5.2 shows a double-decker plot for the variables *Age,*

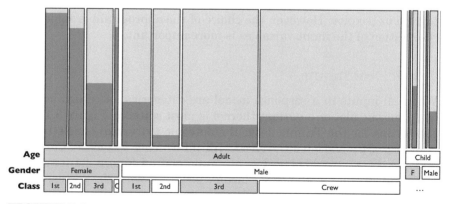

FIGURE 5.2

The **Titanic** data in a double-decker plot, with surviving passengers selected.

Gender and *Class*. In a double-decker plot, all bars are subdivided along x and thus all have equal height and proportional width. The highlighting is done from the bottom up along y, allowing direct comparison across all categories. A drawback of this permutation of the standard mosaic plot is that different categories are difficult to identify without extra labeling. Therefore a key is added at the bottom of the plot to simplify the identification of the classes.

The chosen order in the double-decker plot is crucial for what can be read from the plot. The main variables to compare should be put last in the double-decker plot — *Class* in the example of Figure 5.2 — other factors at a higher level. The variable *Age* was put as first variable to concentrate on adult passengers and declutter the plot (only less than 5% of the passengers were children). Figure 5.2 reveals two new insights (for adult passengers) which are far harder to spot in Figure 5.1. Proportions of surviving passengers are all higher for females than for males, regardless of the class. The expected decline of the survival proportion from first class to second to third cannot be found for males, due to a surprisingly small proportion of males surviving in the second class.

Continuous Inputs

Given many continuous input variables plotting single graphs (such as histograms) for each would use far too much space. Parallel coordinate plots (or parallel boxplots) can efficiently display tens of inputs linked to a response variable. In Figure 5.3 all 8 fatty acids from the olive oil data are displayed in a parallel coordinate plot, and the three level response *Area* is displayed in a barchart. Selecting the cateogory *North* in the barchart

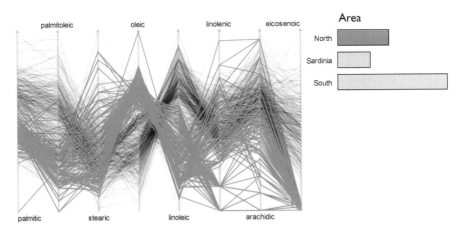

FIGURE 5.3
A continuous response model for the olive oils dataset.

shows which variables might be helpful in classifying the data. Although *eiconsenoic, linoleic* and *oleic* show highlighting only for a small interval, *arachidic* shows highlighting over the whole range of the variable. Furthermore, a highlighted outlier can be spotted in the variable *oleic* which is far off the highlighted group.

The example in Figure 5.3 shows a relatively ideal situation since several of the variables or variable combinations show a clear separation of the highlighted group (it might be necessary to permute the order of the axes and/or invert axes to find the combinations). This might not be the case for other datasets, such that individual scatterplots or more complex visualization methods like Grand Tour and Projection Pursuit might be needed to find potential clusters and groups (if existent).

Mixed Inputs

Sometimes, a dataset features continuous and categorical variables, which are influencing a dependent variable. In such a case, no single multivariate plot can display all variables. Nonetheless, with linking, a combination of plots can still reveal both the dependencies among the influencing variables and their effect on the dependent variable. In Figure 5.4

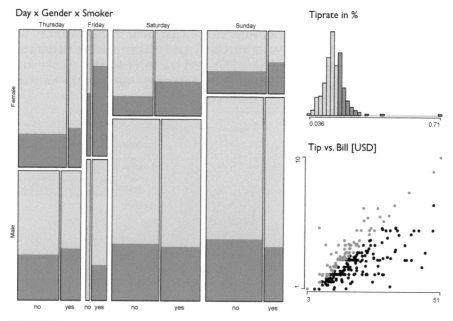

FIGURE 5.4

Five variables which have an effect on the tiprate (response). Three categorical factors in a mosaic plot, two continuous variables in a scatterplot.

a simple example is shown for a mixture of continuous and categorical variables. To study the effect of *Day of Week*, *Gender*, *Smoker*, *Bill in USD* and *Tip in USD* on the actual tiprate, the tiprate is plotted in a histogram with a bin width of 2%, aligned at full percentage breaks. All tiprates higher than 18% have been selected. The scatterplot immediately reveals the trivial linear relationship between tiprate, tip and bill, i.e., Tip = Bill × Tiprate. By querying the point with the highest tip it is possible to see that the tiprate is limited to less than 20% for higher bills, and tiprates above 20% only occur for smaller bills. The mosaic plot shows an increasing number of males paying the bill from Thursdays to Sundays and a hint toward female smokers to be the more generous tippers. For a complete analysis of the tipping data see case study i.

As the example in Figure 5.4 already indicates, with the flexibility of many linked plots displayed simultaneously, there is no fixed set of plots that gives the best results. The most efficient ensemble of plots and variables can be sought, starting with mosaic plots and parallel coordinate plots, in order to understand the data best.

5.2 ANOVA

The influence of one or more categorical factors on a continuous outcome is investigated in a typical ANalysis Of VAriance setting. The question is whether the continuous outcome variable has significantly different values for some levels of a factor or interactions between factors. The one-factor model is usually written as

$$y_{ij} = \mu + \alpha_j + \epsilon_{ij} \quad \sum \alpha_j = 0 \tag{5.1}$$

for $i = 1, \ldots, n$ observations and $j = 1, \ldots, p$ levels of the factor. Adding a second factor will yield

$$y_{ij} = \mu + \alpha_j + \beta_k + (\alpha\beta)_{jk} + \epsilon_{ijk} \tag{5.2}$$

with the constraints

$$\sum \alpha_j = 0, \sum \beta_k = 0, \sum_j (\alpha\beta)_{jk} = 0, \sum_k (\alpha\beta)_{jk} = 0. \tag{5.3}$$

All $\epsilon_{ij(k)}$ are assumed to be normally distributed with mean $\mu = 0$ and a fixed variance σ^2.

A boxplot y by x is the ideal graphical tool to investigate a one-way ANOVA setting. Although ANOVA compares means and boxplots show medians, the overall situation is depicted sufficiently to get an overview

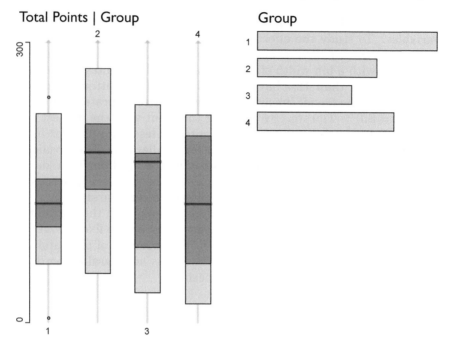

FIGURE 5.5

Boxplot of *Total Points* by *Group* along with a barchart for *Group*.

of the effect of the factor. When medians and means for the groups differ substantially, the underlying distributions more or less violate the necessary assumptions for an ANOVA and the results will be less reliable. Figure 5.5 shows a boxplot of *Total Points* by *Group* along with a barchart for *Group* for the Exams data from case study A. Since boxplots do not indicate the size of the underlying level of the factor, it is a very helpful to display group sizes via a corresponding barchart. There are variations of boxplots which attempt to indicate group sizes by adjusting box width of the boxplot, but those are less powerful in conveying that information than barcharts. Whether or not a difference in means/medians is significant cannot be seen from a boxplot, and depends strongly on the size of the groups. One way to give a rough estimate of whether a group difference is significant is to use so-called notched boxplots. The intervals are constructed such that if two gray boxes fail to overlap, the corresponding medians are discernibly different at approximately 5% significance level. Figure 5.6 shows two implementations of notched boxplots. The left plot shows a notched boxplot in R, the right plot shows the same plot in DataDesk. For the data in Figure 5.5 and Figure 5.6, we find that there is a significant difference between groups 1 and 2.

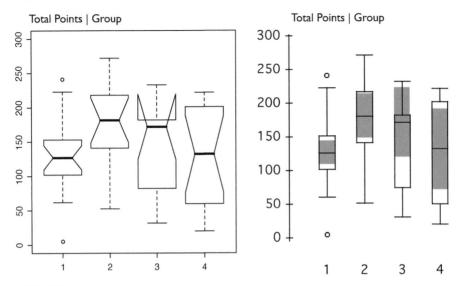

FIGURE 5.6
Notched-boxplot of *Total Points* by *Group*. Left: R, Right: DataDesk

Two-way Interactions

It is obviously difficult to compare boxplots which are not plotted along the same scale. Thus, it is very challenging to investigate a two-way interaction of two factors in a trellis-like grid which has rows and columns. A very simple yet effective plot to look for two-way interactions is the so-called interaction plot.

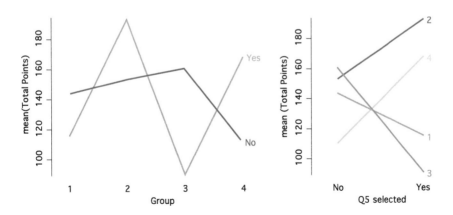

FIGURE 5.7
The two possible interaction plots of *Total Points* ~ *Group* * *Q5 selected*. In both variants non-parallel lines indicate the presence of an interaction.

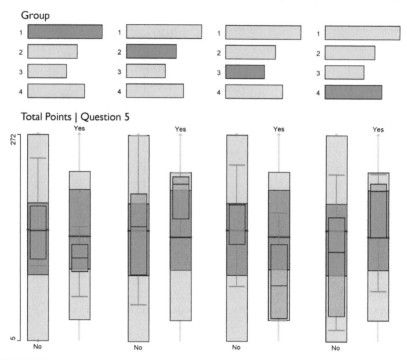

FIGURE 5.8
Traversing the categories of *Group* reveals the interaction via highlighting in the boxplot of *Total Points* by *Q5 selected*.

Figure 5.7 shows the two possible interaction plots of *Total Points* for the two factors *Group* and *Q5 selected*. An interaction plot displays a dot-plot of the expected cell means $\mu+\alpha_j$ and $\mu+\beta_k$ for the specified dependent variable. It displays the levels of one factor on the x-axis and shows the means of each level of the other factor. The y-axis is the dependent variable. Then the expected cell means are connected with lines by group to the other factor of the interaction. If there is no interaction present, all lines are roughly parallel. The more the lines deviate from a parallel pattern, the more this is an indicator for an interaction. In Figure 5.7 (right), none of the lines (each representing a level of the factor *Group*) are parallel — the lines for *Group 1, 3* and *2, 4* are showing inverse trends. In a graphical environment, we want to be able to spot the interaction without calculating the ANOVA and the corresponding expected cell means. This can be achieved by creating a boxplot y by x for the first factor x and a corresponding barchart for the second factor z. This setup is shown in Figure 5.8 for the model *Total Points* $\sim$ *Group * Q5 selected*. The boxplot shows *Total Points* by *Q5 selected*, and the barchart shows the four levels of *Group*. The figure shows four different snapshots for each selected

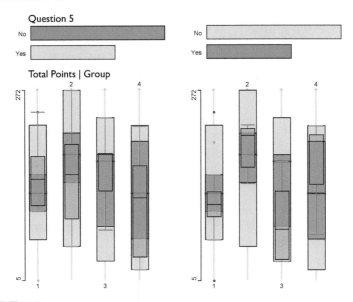

FIGURE 5.9
Exchanging the roles of *Q5 selected* and *Group* makes the interaction even more obvious.

group in the barchart. From left to right, i.e., from *Group 1* to *4*, the highlighted boxplots show an increase for *Total Points* for students who selected question 5 in *Group 2* and *4*, and decreasing values for *Group 1* and *3*. This indicates an interaction, because the effect of the factor *Q5 selected* is not independent of the levels of the factor *Group*. The representation in Figure 5.8 is close to what the interaction plot in Figure 5.7 shows. The fixed assignments in an interaction plot have a different visual impact and are less flexible compared to the perception and possible group comparisons when switching between the levels of *Group* interactively. In the same way in which the role of *Group* and *Q5 selected* could be exchanged in Figure 5.7, the role of the two factors can be exchanged in the boxplot *y* by *x* and the barchart. Figure 5.9 shows the set-up with the role of *Q5 selected* and *Group* exchanged. This version is in fact more efficient to read, since one only needs to switch between two groups, *Yes* and *No*. The highlighting in the two boxplots shows a strong interaction between the two factors because the effect of *Q5 selected* is inverse for half the classes of *Group*. Apparently, students in *Group 1* and *3* were less well prepared for question 5 and unlike students from *Group 2* and *4* could not benefit from answering this question.

The corresponding ANOVA model states that the interaction of *Group* and *Q5 selected* is significant with a p-value of 0.0397. Clearly, graphics cannot give these exact quantities, but the above example shows the benefits of using graphics. The set-up in Figure 5.8 and Figure 5.9 only requires standard plots such as barcharts and boxplots y by x. Selecting the appropriate groups reveals the effect of the two factors and their interaction. Furthermore, using a barchart to select the bars of the respective levels also shows the group sizes.

5.3 Log-linear Models

An example of the extension of statistical graphics with model information was shown in Section 3.3, page 59, where a linear fit or an adaptive smoother was added to a scatterplot.

As already pointed out in Section 4.1, page 74, mosaic plots have the nice property that they can be drawn for both the raw data as well as for any model based on these raw data. Thus, models for categorical data can be displayed in a mosaic plot and compared to the raw data.

A trivial model is the model with all cells having the same size. This is already covered by the same binsize variation of a mosaic plot. The simplest model for categorical data is the model of marginal independence, cf. Figure 4.4.

Log-linear models aim at modeling interactions between more than just two variables. Depending on how many variables are investigated simultaneously and how many interactions are included in the model/data, different model types can be distinguished by simply looking at the corresponding mosaic plot. Each of these models exhibits a specific pattern in a mosaic plot. If there are less than four variables included in the model, the specific interaction-structure of a model can be read from the mosaic plot.

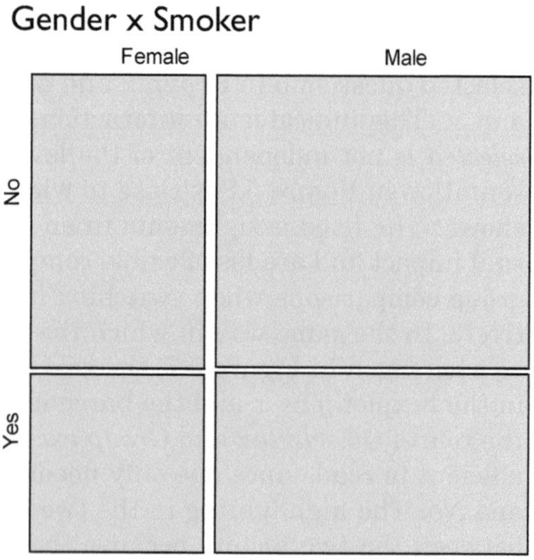

FIGURE 5.10

For the tips data in appendix i *Gender* and *Smoker* are independent.

For the tips dataset from appendix i, we will investigate the relationship between the three variables *Day*, *Smoker* and *Gender* more closely using mosaic plots and log-linear models. In general, for k variables, we can distinguish $2^k + 1$ log-linear models; thus for three categorical variables we can consider nine different models (if we neglect the trivial model of constant cell size). The models are:

1. **Marginal Independence,**
   a model with no interactions present.

2.-4. **Partial Independence,**
   those are the three models with only one interaction between the three variables.

5.-7. **Conditional Independence,**
   the three models where one of the three two-way interactions is missing.

8. **No Three-way Interaction,**
   the model with all two-way interactions present.

9. **Saturated Model,**
   with all two-way and the three-way interactions included.

Usually, we assume the models to be hierarchical: for any $n < m$, the presence of an m-way interaction requires the presence of all n-way interactions. This means that any model from 2.-4. will include 1., any model from 5.-7. will include two models from 2.-4., model 8. includes all models 2.-4. and model 9. includes 8. Thus, when investigating a set of categorical variables for their interaction structure, we will start with looking at all pairs of two-way interactions.

For the above example, we might start by looking at a possible interaction of *Smoker* and *Gender*. The corresponding mosaic plot is shown in Figure 5.10. All separating gaps in Figure 5.10 are almost perfect straight lines which indicates independence. Independence in Figure 5.10 means that no matter which category of *Smoker* we look at, the proportion of *Gender* is always the same. This independence property also holds true when interchanging the role of *Gender* and *Smoker*. Although a two-dimensional mosaic plot usually changes with the order of the two variables, here the independence structure remains the same even if the variables are interchanged. The next pair of variables we look at are *Day* and *Gender*. The corresponding mosaic plot is shown in Figure 5.11 (left). If the two variables were independent, the proportions of *males* and *females* would roughly be the same for all days — obviously this is not the case. Besides the mosaic plot of the raw data, Figure 5.11 (right) also shows a mosaic plot for the same data for the model of mutual independence. In this plot, the gaps are all perfectly aligning, which indicates the independence model. Additionally, the residuals of this model are displayed by blue and red highlighting in the plot. The residual highlighting is done

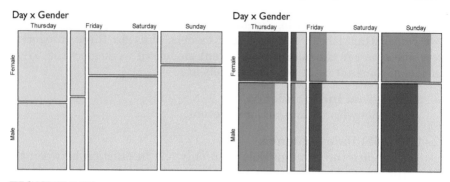

FIGURE 5.11

There is a dependence between the variables *Day* and *Gender*. In the raw data (left plot) we see an increasing number of males toward the weekend. In the modelled data (right plot), assuming independence, the residual highlighting shows an overcrossing highlighting structure corresponding to the linear trend visible in the left plot.

as follows: for the observed counts o_{ijk} and expected values e_{ijk} in each cell we calculate the normalized residuals

$$\tilde{r}_{ijk} := \frac{o_{ijk} - e_{ijk}}{\sqrt{o_{ijk}}} \tag{5.4}$$

as the square root of the contributions of the χ^2 statistics. The highlighting color is chosen by the sign of the $\tilde{r}_{ijk}$; red corresponds to negative residuals, i.e., the model predicts higher counts than actually observed, blue the opposite. The amount of highlighting corresponds to the size of the $\tilde{r}_{ijk}$. The cell with the largest $\tilde{r}_{ijk}$ is completely filled by the residual highlighting, all other cells are scaled accordingly. Besides color and size of the highlighting, the saturation of the highlighting is adjusted according to the model investigated in the plot. For models with a clearly significant p-value (which is set to be a p-value smaller than 0.01), the blue and red highlighting is fully saturated. The saturation is reduced for p-values between 0.01 and 0.10 and becomes very faint for p-values larger than 0.1.*

Figure 5.11 (right) shows the residual highlighting for the model of mutual independence for the variables *Day* and *Gender*. The cell for *females* on *Thursdays* is completely highlighted in blue, indicating that the largest normalized residual is in this group. The trend of an increasing

---

*There are other suggestions of how to code the residual in a mosaic plot. Friendly (1994) proposed plotting the residual information using the mosaic plot for the raw data. This has the drawback that the information is illegible for very small cells and for empty cells not shown at all. Meyer et al. (2007) basically follow Friendly's proposal but use a more refined color shading.

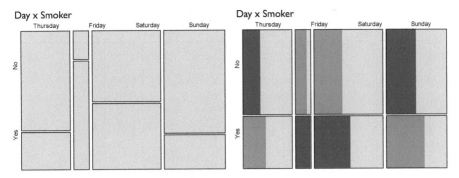

FIGURE 5.12
There is also a dependence for the variables *Day* and *Smoker*. In the raw data (left plot) we see far more smokers on Fridays and Saturdays. In the modelled data (right plot), assuming independence, the strong deviation for *Fridays* is indicated by the highlighting proportion of the residual highlighting.

proportion of *males* toward the weekend is reflected in the overcrossing pattern of blue and red highlighting. The highlighting is plotted in saturated colors indicating that this interaction is highly significant, which matches the impression obtained from the mosaic plot.

Figure 5.12 shows the third two-dimensional mosaic plot, which includes *Day* and *Smoker*. The proportion of smokers is less than 50% for all days except for Fridays, where it is almost 80%. This strong deviation from the general trend is also visible in the corresponding mosaic plot for mutual independence with residual highlighting. The largest normalized residuals can be found in the two cells for Fridays. The relatively small highlighting proportions for Thursdays indicate that the average proportion of smokers vs. non-smokers is close to the proportion found for that day. The strong saturation of the highlighting shows that this interaction is also highly significant.

Figures 5.10 to 5.12 show all three two-way plots for the three variables. One of them, *Gender* x *Smoker*, did not show an interaction, the other two, *Day* x *Gender* and *Day* x *Smoker*, showed interactions. Since we are interested in a model for all three variables, we need to look a mosaic plot for the variables *Day*, *Smoker* and *Gender*, which is shown in Figure 5.13. From examining the two-way interactions we learned that including one two-way interaction will most likely result in a signifiacnt model.

Figure 5.14 shows the stepwise construction of a log-linear model for the three variables *Day*, *Smoker* and *Gender*. The topmost plot shows the model of mutual independence, which is highly significant. Adding the interaction between *Day* and *Smoker* results in a model of partial

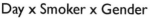

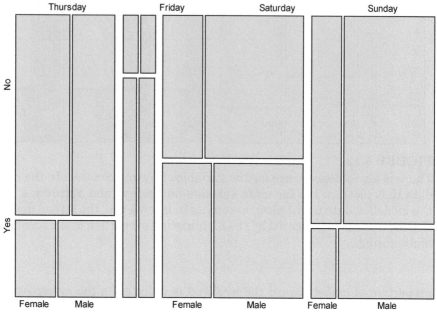

FIGURE 5.13

The mosaic plot for all three variables *Day*, *Smoker* and *Gender*. Given a level of *Day*, *Gender* and *Smoking* still appear to be independent.

independence. The middle plot in Figure 5.14 shows the modelled data with the interaction between *Day* and *Gender* included. Because these two variables are the first in the plot, the interaction structure is directly visible. The saturated color of the residual highlighting indicates a still significant deviation of the model of partial independence from the raw data.

Looking at the residual highlighting in the middle mosaic plot of Figure 5.14 shows blue highlighting for males on Saturdays and Sundays and red highlighting for Thursdays and Fridays (vice versa red highlighting for females on the same days). This is an indication of an interaction between *Day* and *Gender* which still has to be added to the model. Note that interactions can be best seen between the first two and the last two variables of a mosaic plot.

The lower plot in Figure 5.14 shows the final model of conditional independence, which includes the interactions between *Day* and *Smoker* as well as *Day* and *Gender*. In this model, *Smoker* and *Gender* are independent given a level of *Day*. The ratios of smokers/non-smokers and females/males are different for different levels of *Day*. Comparing this model with the raw data in Figure 5.13 shows only a slight deviation,

Day x Smoker x Gender

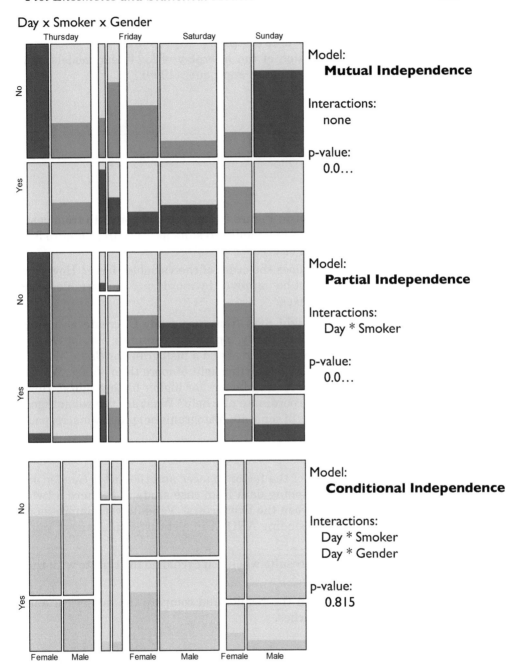

FIGURE 5.14
Stepwise development of a log-linear model for the variables *Day*, *Smoker* and *Gender*. The residual highlighting indicates size and sign of the residuals, the saturation shows the overall significance of the model.

which is also supported by the very faint color of the residual highlighting corresponding to the p-value of 0.815.

A more detailed discussion of the interplay of log-linear models and mosaic plots can be found in Theus and Lauer (1999).

Exercises

5.1. Response Models

    (a) Recreate the plot in Figure 5.3 for the olive oil data from case study H. Can you find similar results as those from Figure 5.3 for other regions or areas?

        What influence does the order of the variables have? How can the visual impact be improved by reordering the variables and inverting their axes?

    (b) For the birthweight data from case study C, create a mosaic plot for *Smoker* and *Parity*, a parallel coordinate plot for *Gestation, Height, Weight* and *Age* and a histogram for *Birth Weight*. Select all cases with a birthweight of more than 3500g. Which are the most influencing factors for higher birthweights? How can the parallel coordinate plot help? What are the advantages or disadvantages of separate histograms or spinograms, respectively?

5.2. ANOVA

Check the influence of the factor *Smoker* and the factor *Gender* on the tiprate for the tipping data from case study i. Is there a two-way interaction between the two factors? Validate your impression by running a corresponding ANOVA in a statistics package of your choice.

Do you get the same results when you exchange the tiprate with the tip in USD?

Interpret and explain the results and compare the numerical and the graphical approaches.

5.3. Log-linear Models

The following four mosaic plots show an unspecified dataset consisting of three binary variables.

The four plots include:

    1. the raw data

2. the independence model
3. the model of partial independence
4. the model of conditional independence

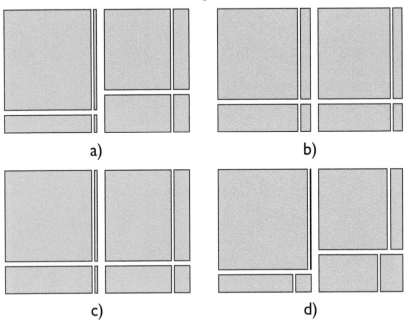

a)   b)

c)   d)

Discuss how the different models can be identified in a mosaic plot for the three variables. Try to assign the four plots a–d to the different models 1–4 by comparing their properties.

6

Geographical Data

Whenever the cases of a dataset can be related to geographical locations, this extra information is a valuable addition that must not be neglected. Geographical data may come in many flavors. The most common are

- **Points**

  If the geographical reference is simple coordinates, i.e., longitude and latitude, the geographical aspect can be plotted in a scatterplot. In this case, the geographical part of the data can be handled with standard statistical graphics. Figure 6.1 shows the locations of

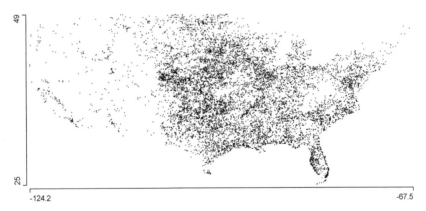

FIGURE 6.1
A scatterplot of the locations of almost 11,000 measurements of tornadoes in the continental U.S. between 1992 and 2001.

10,997 measurements of tornadoes in the continental U.S. between 1992 and 2001 in a scatterplot.

- **Paths and Flows**

  The data in Figure 6.1 are already an example of an entity (in this case a tornado) contributing several observations to the dataset. If the data describe flows such as transportation of raw materials, traffic into a city or the paths of hurricanes, the minimal representation of such a graph is a polyline. Figure 6.2 shows an example of hurri-

FIGURE 6.2
Paths of hurricanes registered in the year 1967 (red). All points where measurements of hurricanes were recorded between 1945 and 1979 are marked as dots.

cane data in the Atlantic and the Gulf of Mexico over the years 1945 to 1979. The paths of the hurricanes in the year 1967 are highlighted. Additionally, all points where measurements were taken are plotted in the coordinate system.

- **Polygons/Choropleth Maps**
  Whereas the two representations of geographical information above can be displayed using a scatterplot, maps built up by polygons require a plot type of their own. Figure 6.3 shows a map of the 25

FIGURE 6.3
A map of the 25 districts of the city of Munich.

districts of the city of Munich from case study E. Maps can be seen as glyph-based plots, though the glyph has an individual and often very complex shape. Within a selection, each polygon that intersects the selection area is selected. A selected polygon is entirely highlighted.

- **Raster Maps**
  Raster maps — often also called raster images — represent measurements on a regular grid. They are usually a result of remote sensing techniques via satellites or airborne surveillance systems. They fit neither the construct of scatterplots nor that of maps. Nevertheless, both scatterplots and maps can be used to display raster maps within statistics software which has no extra GIS capabilities. Figure 6.4 shows an example of a raster map displaying the altitude of an $x \times y$ grid of the English Lake District.

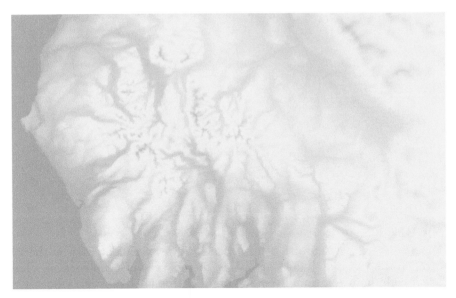

FIGURE 6.4
Raster image of English Lake District.

As can be seen from the first two examples in this chapter, none of the case studies in this book has its primary focus on geographical issues. Thus, the datasets used in this chapter are not necessarily linked to a case study in the second part of the book.

Choropleth Maps

The most common way to display geographically referenced data is choropleth maps. Even if a finer resolution for the geographical location of the items can be given, the interest lies mostly in the aggregated data represented by administrative entities such as ZIP-code areas, NPA-NXX-areas, Counties or States. These areas are then displayed in maps, usu-

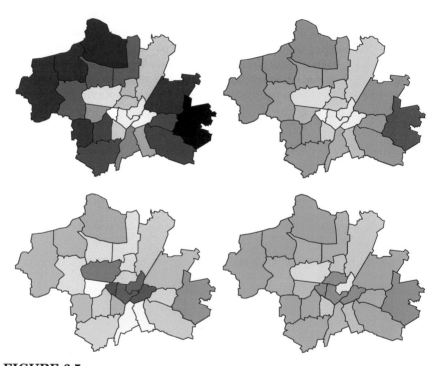

FIGURE 6.5
Four different ways to use color schemes in a map. The map of the 25 Munich districts is shaded according to the year of construction. The lower right map additionally shows highlighting of all districts where less than 85% of the apartments have central heating.

ally shaded according to certain attributes. These (color) shaded maps are called choropleth maps. Figure 6.5 shows examples of four different color schemes. The upper left plot shows a gray-scale shading, the upper right a corresponding monochrome shading in red. The lower left plot shows a diverging scheme using two colors. The lower right plot gives an example where color shading and highlighting is used in the same map. As for all applications where color is used apart from the highlighting, it is imperative to choose the color scheme carefully. Depending on the quantity displayed in a map, the one or the other color scheme is more appropriate. For instance, a diverging blue to red color scheme can be used for temperatures or the percentage of republican votes in a U.S. election.

Visual distortion of Choropleth Maps by Glyph Boundaries

In a map based on polygons, we are accustomed to seeing the boundaries of the areas plotted as solid black lines. For a gray-scale shaded map it leads to an effect of dark shaded polygons being visually enlarged by their boundary. Choosing white boundaries only reverses the effect such that lightly shaded areas benefit visually from their boundaries. One solution

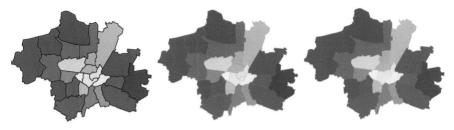

FIGURE 6.6
Three different opacities of glyph boundaries: Left 100%, middle 25%, right 0%.

to reduce this effect is to make the boundaries transparent to a certain degree. Figure 6.6 shows the map of the 25 Munich districts with three different settings for the transparency of the boundaries. Depending on the map, one wants to seek the degree of transparency which still allows us to discriminate the areas while not over-representing darker areas.

Although the effect is less pronounced for the map in Figure 6.6, it is more dramatic for maps with many small areas such as for the East-Cost Counties of the County map in Figures 6.9 to 6.11.

To leverage fully the benefit of different settings for the area boundary transparency, an interactive change of the transparency with a fast update of the map will help to find the right setting for a specific map.

Defining Color Mappings in Choropleth Maps

The visual representation of a choropleth map depends strongly on the way data values are mapped to the chosen color scheme. We use a dataset consisting of socio-economic attributes for the over 3000 U.S. continental counties as an illustration of the different options. The variable we use for color shading is *Males per 100 Females*. Figure 6.7 shows the distribution of this variable. Most observations spread around 100 in an interval from 80 to 120, but 35 counties — just about 1% of the data — have values far higher than 120.

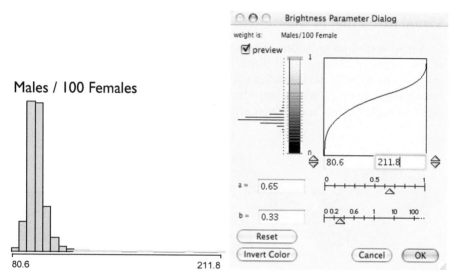

FIGURE 6.7
Left: The distribution of *Males per 100 Females*; Right: The dialog-box for changing the gray-scale mapping in MANET.

Due to the skewness of the distribution, assigning color shades linearly to values of the variables will assign most of the values to very similar color shades. Only outliers will stick out with another color (compare Figure 6.9 top). The corresponding histogram of the upper panel of Figure 6.9 shows the distribution of the shades used. To overcome this problem, Unwin and Hofmann (1988) suggest an S-shaped transfer function to map values to shades using two parameters, a and b. The function is defined as:

$$f(x) := \begin{cases} a \cdot \left(\frac{x}{a}\right)^b & \text{for } x \leq a \\ 1 - (1-a) \cdot \left(\frac{1-x}{1-a}\right)^b & \text{for } x \geq a \end{cases} \quad \text{for } a \in [0,1] \text{ and } b > 0 \quad (6.1)$$

The parameter a determines the centering and b determines the form. Figure 6.7 shows an implementation of the transfer function's dialog box in the MANET software. The parameters can be set via sliders and the corresponding transfer function is displayed along with a draft histogram of the output distribution of the transformed data. Although the parameters were chosen to use the full scale of the shading better, most of the data is still distributed over a relatively small interval.

Choropleth maps are most effective when the range of the color-shading is fully used, i.e., the visual discrimination is maximized. A skewed distribution such as the one shown in Figure 6.7 will shrink the chosen colors to just a fraction of the possible color range. Using a continuously differentiable transformation function such as the one in equation 6.1 is one way to expand the range of colors used. A more effective way to maximize the visual discrimination in a choropleth map is to transform the data to match a target distribution. One option is to force all colors to have the same frequency, i.e., to force the target distribution to be uniform. Another option is to force a normal target distribution. Obviously, the transfer function needed for this transformation is data dependent and piecewise linear. Figure 6.8 shows three different transfer functions to match

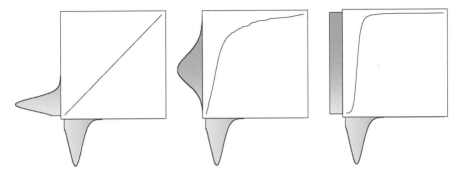

FIGURE 6.8
The three transfer functions that transform the input to identity (left), normality (middle) and uniformity (right).

different target distributions of the color shadings. The left transfer function in Figure 6.8 is the identity function, matching values to color shades linearly. The corresponding choropleth map is shown in Figure 6.9 top. The few outliers dominate the map with a few darker Counties, the vast majority of the other Counties do not show much contrast and are almost indiscernible. The middle transfer function forces normally distributed color shades. The corresponding choropleth map in Figure 6.9 middle has far more contrast, and allows a better discrimination of the Counties. The right transfer function maps the color shades to a uniform distribution, thus maximizing the contrast of the corresponding map, shown in Figure 6.9 bottom.

Comparing the three maps on Figure 6.9 shows the advantage of maps that use a predefined target distribution of the shades over the untrans-

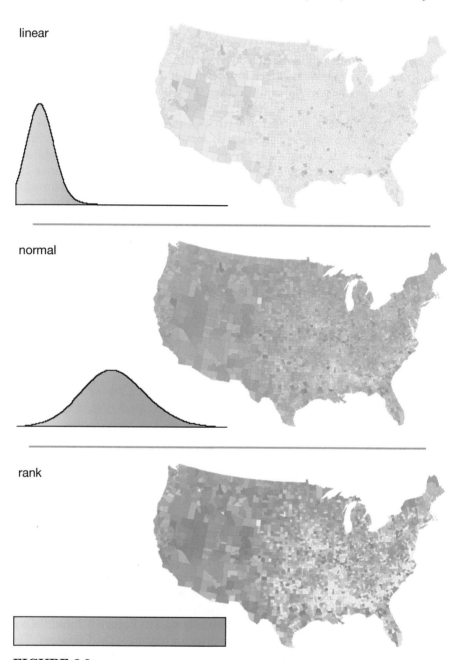

FIGURE 6.9

Three different possibilities to map the quantity *Males / 100 Females* in a choropleth map. Top: linear mapping, Middle: normal distributed colors, Bottom: uniform distributed colors.

formed data. Usually the normalized color shades will give the desired discrimination of areas. If an even higher contrast is needed, the uniform distribution — which in fact assigns the color shades by rank — can be used. When creating a color shading based on target distributions, ties in the data must be mapped to the same color shade, in order to ensure a proper mapping of values to color shades.

In an exploratory and interactive set-up, choropleth maps are usually not used to read quantitative information from the map — this can be easily and far more effectively done by queries — but to get qualitative results which highlight structural features in the data. Therefore, none of the maps showed a scale for the color range. Furthermore, the quantitative interpretation of choropleth maps is intricate, because a value twice as big will not result in a "twice as dark" region.

Outliers in Choropleth Maps

It is often difficult to distinguish between a skewed distribution and an approximately symmetrical distribution with a number of outliers at one end of the scale. In the example of *Males / 100 Females* we found 35 Counties which have far higher values not falling into the range from 80 to 120. These outliers usually require a special treatment during an exploration anyway, such that it is not desirable to give them too much visual weight.

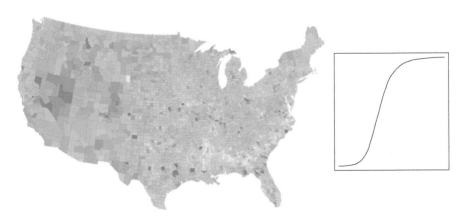

FIGURE 6.10
A linear mapping of values to color shades can still be satisfying when the lowest and/or highest displayed value is limited. Limiting *Males / 100 Females* to a value of 120 is shown in the above map. The cumulative distribution function (right) shows almost normality.

Figure 6.10 shows the choropleth map for *Males / 100 Females* with a linear transfer function, but all values that exceed 120 are limited to that value. The results are very close to the map shown in Figure 6.9 middle. The cumulative distribution function (Figure 6.10 right) shows almost normality of the distribution when the 35 most extreme values are limited to 120.

The advantage of such a solution over a non-continuously differentiable mapping is the better interpretability of the color range used, which is especially useful when a legend must be supplied to decode the color scale in presentation form.

Binary Maps

A binary map is a choropleth map which shows only two states. It can either be drawn for a given binary variable, or for a continuous variable with a defined cut-off value c. All values below the cut-off are assigned the one color, all above the other.

In the example of Figure 6.11, $c = 100$ is the natural cut-off, which corresponds to an exact balance of female and male population.

With the ability to limit the range over which data values are assigned to colors, a binary map may be drawn without the need to derive a binary variable. For a chosen cut-off c, the minimum and the maximum of the color range can be set very close around c, such that all values are either smaller or bigger than the chosen minimum and maximum; thus no values fall within the interval and all values are either assigned the minimal color value or the maximal color value.

FIGURE 6.11
A binary map for *Males / 100 Females* with a cut-off value of 100, showing regions of male and female surplus in County population.

As the data in Figure 6.11 is rounded to two decimals, limiting the range to, e.g., 99.995 to 100 or 100 to 10.005 will do the job.

This chapter covers only important issues associated with using choropleth maps during an exploratory analysis in an interactive environment. Many other options can be used to define a choropleth map. In contrast to classical cartography, the aim is not to create the "perfect" map representing the data, but to provide many different views that will reveal as many interesting aspects of the data as possible. The flexibility of the interactive tool is the key feature to ensure an efficient work-flow.

Mondrian includes almost all of the features for choropleth maps described in this chapter and offers tight integration with traditional statistical graphics.

Exercises

6.1. Create the scatterplot from Figure 6.1 using the data for longitude and latitude in the *tornadoes* dataset (available from the book's website).

    (a) How can different plot sizes, point sizes and α-levels show regions of high tornado activity?

    (b) Select the points from different months. Is any geographical pattern visible?

6.2. Create the same map as in Figure 6.4 lower right for the Munich rent data from case study E. Use a histogram of *Built* set to bins of full decades to color-brush the map. Derive a discrete variable from this color mapping and use this variable to create a choropleth map. Compare all three maps, and discuss their advantages and drawbacks.

6.3. In Figure 6.11 a cut-off of 100 was chosen under the assumption that the two genders occur equally frequently. Find the current proportions of females and males in the U.S. from the US Bureau of Census. Set the thresholds of the map to reflect this ratio. Which c do you need to chose? Does your map differ substantially from the one in Figure 6.11?

7

More Interactivity

The basic concepts of interacting with graphics such as selections and queries have been introduced in Chapter 1. There are far more interactions, mostly changing the parametrization of a plot. Several more plot specific interactions have been used in the previous chapter without explicitly mentioning them. Switching a barchart to a spineplot (compare Figure 3.1), a histogram to a spinogram (compare Figure 3.6), or changing a mosaic plot from the default view to the model of independence (see Figure 4.4) are all interactions which should be just a single click or keystroke away, without the need to recreate the "same" plot with different plot parameters.

These interactions are often very plot specific and thus cannot be discussed in a general context. Other interactions such as sorting, zooming and the creation of multiple views refer to more general concepts and are investigated in this chapter. Not all of the techniques mentioned in this chapter is available in Mondrian or other software yet, but the broader context will help us use the already existing tools efficiently.

7.1 Sorting and Ordering

Sorting data is one of the most efficient actions to derive different views of data in order to see the variables from many angles. Sorting is usually not applied to the data itself, but to statistical objects of a plot. We might want to sort the bars in a barchart, the variables in a parallel boxplot or the categories in a boxplot y by x.

In principle, sorting operations can be categorized into automatic sorting options and manual sorting options. Manual ordering facilities are most helpful when the number of objects to sort is small, visual feedback is needed, or no quantity is available that determines the order. As soon as the number of objects grows, automated sorting facilities should be provided.

Sorting Categories in a Barchart

There are two sorting orders for barcharts which can be applied regardless of the selection state in the dataset. One can sort by

1. frequency, or

2. category name.

Two more sorting options can be considered when the selection state is taken into account. Sort by

3. absolute highlighting

4. relative highlighting.

Figure 7.1 shows a barchart of the variable *Type of Rider* for the Tour de France 2005 data (cf. case study F). In the upper left barchart the default sorting is shown using the order according to the level names of the variables. The upper right barchart is sorted according to the frequencies, which immediately shows that *Climber* is the smallest group and *Helper*

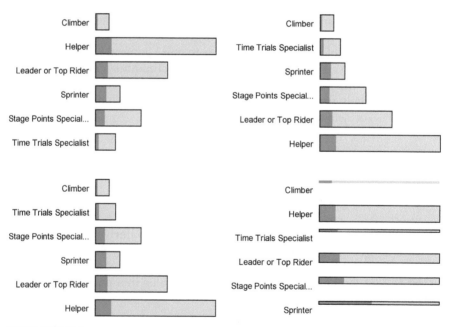

FIGURE 7.1

Four different sorting orders for the barchart of *Type of Rider*. Top left: lexicographic order, Top right: Sorted by category size, Lower left: Sorted by absolute highlighted, Lower right: sorted by relative highlighted. (All riders who dropped out of the Tour de France 2005 are highlighted.)

is the largest group. So far the highlighting information was not used to sort the categories. The highlighting in Figure 7.1 corresponds to the 34 riders who did not finish the Tour for whatever reason. The lower left barchart is sorted according to the absolute number of selected cases. The categories of *Sprinter* and *Stage Point Specialists* have changed places compared to the upper right sorting, indicating that there is no independence between the two attributes. Switching to a spineplot sorted according to the relative number of selected cases per level shows a clearer picture. Almost 50% of sprinters could not finish the race, whereas only about 10% of all climbers dropped out.

For variables with very many levels — which cannot be displayed in a barchart without scrolling — it is helpful to have the possibility of inverting the sorting order in order to achieve ascending and descending order for all sorting criteria.

Manual sorting is indispensable in all cases where the sorting cannot be derived from information within the dataset. Figure 7.2 shows the four weekdays from the tipping data in case study i in default order. Obviously *Thursday* should be at the first position, not at the last position. Since none of the four automatic sorting options listed above will yield the desired order of the four weekdays, a manual ordering by dragging the bar to the correct position has to be performed.

No matter how a sorting order has been defined, the new ordering of categories must be propagated to all plots which use this variable, be it a mosaic plot, a boxplot y by x, or a choropleth map. This is another form of linking, which does not link the selection attribute, but links the axis attribute of the categorical variable. The axis information of a categorical variable is basically the order of its categories. For a continuous variable the minimal axis information is the range to be displayed. Linking of continuous axes is usually optional and results in plots of the same scale. Identically scaled plots are very important for comparing values across different plots.

FIGURE 7.2
A simple drag and drop operation can be used in order to sort levels to the desired position manually.

Reorder Variables in High-Dimensional Plots

Although mosaic plots and parallel coordinate plots are of very different type, ordering their variables to a sensible order is essential for their usability. The more easily these orderings can be modified interactively, the more effective these plots are. The benefits of different ordering in mosaic plots and parallel coordinate plots have already been discussed on pages 72 and 78 respectively.

7.2 Zooming

Zooming operations become increasingly important with the size of a dataset. Switching between overview and detail for smaller datasets usually makes no big difference — for larger datasets, it is often essential.

Standard zoom operations result in a "simple" change of the scale of the axes involved. Figure 7.3 shows an example of a sequence of zoom operations in a scatterplot. For the Augsburg data from case study G a scatterplot for the tax payment in 1646 vs. the tax payment in 1618 is plotted. The upper left scatterplot shows the complete data, un-zoomed. The plot is dominated by one outlier, which has twice as many payments as all other districts. The first zoom step reduces both axes by a factor of 10, thus showing only a 100th of the original scale, still covering roughly 80% of all data points. The next zoom step (lower left plot in Figure 7.3) reduces the scale by a factor of 100 again. This scatterplot still covers almost half of all data. Whereas the first zoom reveals the self similarity of the upper left and the upper right plot, the second zoom shows that there are also a few smaller districts which did not suffer a decline in tax payments over the course of the war.

The lower right panel finally visualizes the two zoom steps in relation to the original plot area. The last zoom, being only a 10,000th of the original size, is almost invisible. For the Augsburg data, we are looking only at 95 data points. Larger datasets, such as tax data or data on financial transactions, face far more extreme zoom operations, where a very few outliers expand the scale by several orders of magnitude (cf. Unwin et al. (2006) Chapters 3 and 4).

Zooming into a scatterplot as shown should be possible by interactively selecting the zoom-region or by explicitly specifying the new bounds of the plot. When zooming in and out of a graphic, be it a scatterplot or a map, it is important for a zoom-out step not to expand the axes by a constant factor arbitrarily. Instead, it should step back to the last zoom step such that the user can more easily preserve the context as much as possible.

An alternate approach to handling data such as the one shown in Fig-

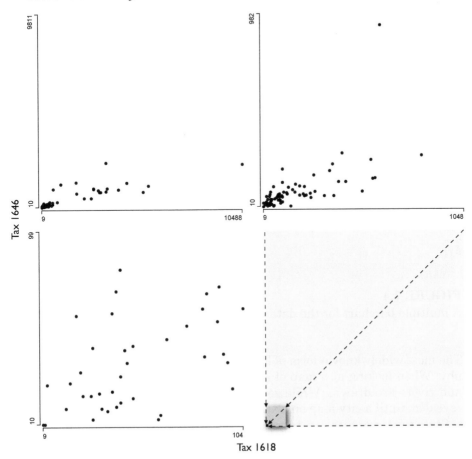

FIGURE 7.3

Two zoom operations (from top left to bottom left), each zooming in by a factor of 100, reducing the initial scale by a total factor of 10,000. The lower right panel visualizes the relations of the three chosen scales.

ure 7.3 would be to use a log-transformation of the data. Zooming supports Shneiderman's (1996) visual information-seeking mantra "Overview first, zoom and filter, then details-on-demand." The static solution of a log-transformed scatterplot can neither reveal an overview of the data in the original domain, nor show specific details, but might still be appropriate if the data must be presented in a static context.

Logical Zooming

Logical zooming covers all kinds of zooming, where the representation of the data changes depending on how much detail of the data is shown.

Marital Status x Region x No. of Children

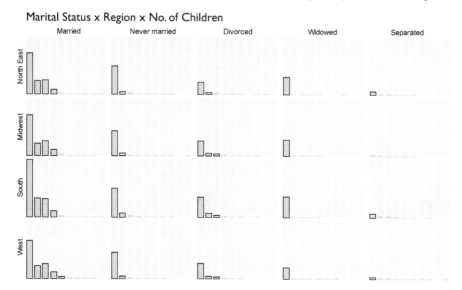

FIGURE 7.4

A multiple barchart for the data from the *95 Current Population Survey.*

The most widely known form of logical zooming can be found in cartography. When looking at a map of a whole country, only major cities, roads and rivers are drawn. While zooming in, more and more detail will be revealed, until a city map on a street level might be reached.

Statistical graphics can borrow from these techniques. Geographical data may be stored on different levels of aggregation (State and County); scatterplots of massive data may be binned* and only reveal raw data once the user zooms in deep enough to show the detailed view.

Censored Zooming

Whenever large differences occur between the biggest and smallest values zooming is an effective way to cope with these data. For continuous data, classical zooming can help to switch between overview and detail. Area- based plots depicting categorical data do not necessarily support such types of zooming operations. For area-based plots, the so-called censored zooming can be applied. Censored zooming comes in two versions, floor-censored zooming and ceiling-censored zooming, the latter being needed more commonly. Zooming-in increases the size of objects, zooming-out decreases the size of objects. Ceiling-censored zooming works by censoring (limiting) the size of objects at the size of largest ob-

---

*A detailled discussion on binned scatterplots can be found in Chapter 9.

Marital Status x Region x No. of Children

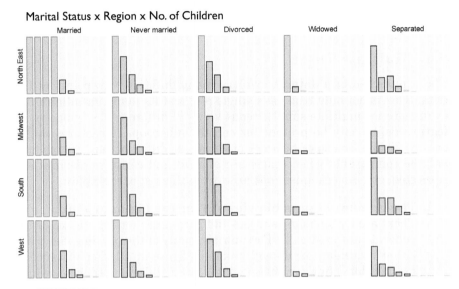

FIGURE 7.5

A censor-zoomed multiple barchart for the data from the *95 Current Population Survey*. Censored cells exceeding the maximum size are marked by a red frame.

ject of the un-zoomed plot. When zooming out, objects cannot get smaller than their original size. Analogously, floor-censored zooming defines the minimal size of an object which is needed in order for it to be drawn. All objects smaller than this threshold are not drawn.

For barcharts, censored zooming can be emulated by simply limiting the minimum and maximum of the count axis. For mosaic plots, the definition of the plot must be extended.

Figure 7.4 shows the multiple barchart for the variables *Marital Status* x *Region* x *No. of Children* for the *95 Current Population Survey* data. The distribution of *No. of Children* can easily be seen for the category *Married* but is almost invisible for the other categories of *Marital Status*. Figure 7.5 shows the corresponding ceiling-censored zoom of Figure 7.4. The zoom has been performed such that the largest bar for the category *Separated* completely fills its rectangle. The distribution of *No. of Children* can now be seen even for the less populated levels of *Marital Status*, and the second mode for 2-children households which was visible for the category *Married* can only be found for *Separated* in the *North East*.

7.3   Multiple Views

The exploration of a dataset involves many different viewpoints of the data. This is not only true for a single plot and its many variations, but even more for multiple simultaneous views in different plot windows. A feature might only be present for certain subgroups, structures might only be visible for a specific order of variables, differences of groups may only be comparable when scales are aligned and certain orderings are respected, etc. Several questions arise from this list of examples:

1. **The number of possible views is far too big to explore.**
   There are far too many subsets and parameter settings to explore or generate automatically. Fortunately, many of the possible views are either redundant or do not make any sense at all. For instance, a mosaic plot of five variables yields $5! = 120$ different orderings which is impossible to look at even with software support. In contrast, a parallel coordinate plot with 5 variables requires only 3 permutations to show all possible adjacencies when systematically generated by the software.
   How can interactive software help guide an exploration and make the most promising views readily available?

2. **Default settings for statistical graphics are crucial.**
   Graphics (multivariate in particular) have many degrees of freedom regarding their initial parametrization. Only a few distinct settings may reveal interesting information in the data. For instance, the α-level and the size of the points in a scatterplot can be set automatically depending on the expected amount of overplotting.
   How can default settings of statistical graphics be improved in order to maximize their possible information content?

3. **Managing multiple views is not trivial at all.**
   With the flexibility of an interactive software system at hand, many different views can be generated very quickly. Arranging windows and matching plot parameters is then no longer a trivial task. For instance, it is easy to open 10 different histograms of variables measured on the same scale. If it was necessary to scale all of them manually in order to have the same x-axis and y-axis, one would probably leave out such a comparison.
   How can the software aid the user to manage multiple views effectively, without losing the flexibility needed for interactive graphics?

None of the points mentioned above can sufficiently be answered in general. Nevertheless, software should respect their implications as much as possible. Many of the interactions described and used in the previous chapter address the above issues partially. Given the possibilities of today's software users should definitely ask for more. Features such as an effective implementation of simultaneous plot windows or a sensible arrangement of plot windows on the computer screen when either opened at once or sequentially seem to be basic tasks but are far from being commonplace.

7.4 Interactive Graphics ≠ Dynamic Graphics

The two terms "Interactive Graphics" and "Dynamic Graphics" are often mixed up. Although interactive graphics can be seen as a superset in which dynamic graphics is only one of many techniques a more detailed discrimination can be given.

Dynamic graphics subsumes all techniques where one or more parameters of a plot are changed continuously and the corresponding plots update smoothly. The most prominent example is a 3-d rotating plot, where the projection parameters are changed steadily to show a smooth pseudo 3-d rotation of a point cloud. Other techniques which are often used are interactive changes of transformation parameters, which can then be monitored in all associated linked plots.

Historically, dynamic graphics techniques were the first interactive statistical graphics to be implemented in software tools. Watching a 3-d rotating plot on your desktop computer was quite fascinating in the early to mid-1980s. In the long run, dynamic techniques are only one part of a far broader spectrum of interactive graphical tools for data analyses.

The most interesting dynamic graphical tools that prevailed are the Grand Tour and Projection Pursuit Methods.

Grand Tour

The Grand Tour is a generalization of a 3-d rotating plot. A computer screen can display 2-dimensional objects only. A 3-d rotating plot is the projection of a 3-d object, e.g., a point cloud, onto the 2-d screen. The pseudo rotation is achieved by displaying a rapid succession of projections with only a small difference in their parameters.

Since we live in a 3-dimensional world, our visual system is able to interpret the pseudo-rotation in a suitable way, although it is only presented as 2-dimensional information. Using projections, there is no reason why we cannot look at point clouds in four or even higher dimensions. The only problem is the way the rotation can be controlled and the rotated object interpreted. Three natural axes — x, y and z — exist for a 3-d rotation. It is a natural task for the user to rotate an object around one of these three axes. Beyond three dimensions the Grand Tour is needed as an automatic steering mechanism of the rotation. It is formally defined as:

> A continuous 1-parameter family of d-dimensional projections of p-dimensional data which is dense in the set of all d-dimensional projections in $I\!\!R^p$. The parameter is usually thought of as time.

Although this definition might sound quite technical, it can be easily explained when looking at a 3-d rotating plot. In a 3-d rotating plot, the projection is of dimension $d = 2$, the computer screen, the data is of dimension $p = 3$. Dense means that the Grand Tour randomly selects projection planes in $I\!\!R^p$ in such a way that the whole space is well covered. The transition between two successive projection planes is interpolated such that the pseudo-rotation appears to be smooth.

For objects of dimensionality 4 or higher, we are not well trained to understand the meaning of a rotation around the axes z_1, z_2, z_3 and z_4. Generally we do not have an understanding of how an object looks in more than 3 dimensions.

The unit cubes for dimensions 2, 3 and 4 are displayed in Figure 7.6. A unit cube of dimension 2 is obviously just a square. For dimension $p = 3$, we get what we usually would call a cube. For dimension 4, most people will not recognize the object as a cube. But what do rotations of these unit cubes look like? For $p = 2$ it will be a rotation of the square around its center. For $p = 3$ the rotation will be easy to understand, as we know how to visually assemble the 6 rhombuses to a cube. The rotation in 4-d is quite different. Although the 24 faces are still just rhombuses in the projection, it is beyond our imagination and thus we cannot assemble it into a known structure. Similar problems will arise when trying to interpret the structure of data in more than three dimensions and much training will be needed to understand what we see.

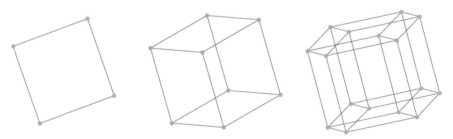

FIGURE 7.6
The unit cube in 2, 3 and 4 dimensions.

Nonetheless, when rotating high-dimensional point clouds of continuous data with the Grand Tour, it is often possible to spot interesting projections which reveal groups, holes, outliers or other structures in the data that might not be visible in lower-dimensional projections.

Figure 7.7 shows a sample snapshot of a Grand Tour in a ggobi session. The example shows all 8 continuous variables from the olive oil data from case study H. The snapshot was taken for a projection which separates the three groups color brushed according to the 3 areas shown in the linked barchart in the lower right.

Projection Pursuit

Although the Grand Tour allows us to look at high-dimensional data, just watching the rotation and hoping to see an interesting projection may not

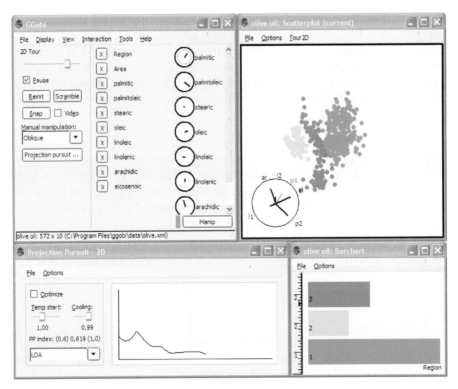

FIGURE 7.7
A ggobi session. The upper left window holds the variable list, the rotational controls and most of the menus. The upper right plot shows the toured data; the lower left window the Projection Pursuit controls and the lower right plot the group assignment in a barchart

necessarily give satisfying results and may take too long to be feasible. At this point Projection Pursuit can be used to systematically search $I\!R^p$ for features. At the core of Projection Pursuit is the so-called projection index. A projection index measures a specific feature such as separation of groups, central mass or holes of the current projection. An optimization algorithm then determines in which direction the data should be rotated in order to increase the projection pursuit index. Obviously, the optimization must find a good trade-off between easy to find local maxima and a probably impossible to determine global maximum.

Figure 7.7 shows an example of the technique of Projection Pursuit implemented in the ggobi software. In the upper left, the main control window shows the Grand Tour controls and the variable circles which indicate the current projections of the variables in the Grand Tour. The upper right scatterplot shows the current projection of the data. The lower right barchart shows the group assignment. The window on the lower left is used to control and monitor the Projection Pursuit. Starting the optimization, the Grand Tour will stop to randomly select the next projection. In a Projection Pursuit the new projection is chosen to optimize the projection index instead. The chances of finding the global optimum are fairly small unless the dimensionality is low. Nonetheless, the interactive nature of Grand Tour and Projection Pursuit allows us to test many different starting values for the optimizer, which will hopefully lead to different solutions.

Both dynamic graphics techniques, Grand Tour and Projection Pursuit are only usable in an interactive implementation. The ggobi package (http://www.ggobi.org) implements these techniques, and allows us to explore continuous data using dynamic graphical techniques.

Exercises

7.1. Ordering

   Select all apartments which cost more than €10 per square meter in a histogram for the rental data from case study E and sort the barchart of the districts according to

   (a) absolute highlighting

   (b) relative highlighting.

   How do the sort orders differ?

Sort the districts by their median price per square meter using a boxplot y by x.

7.2. Zooming

Create the same series of plots as in Figure 7.3, but make sure that the scales are such that the bisecting line of an angle separating growing and shrinking Counties has an angle of 45°. Does this change the interpretability of the plots?

7.3. Multiple views

Create a histogram of the team time trials at stage 4 and the time trial at stage 20 of the Tour de France 2006 data from case study F. Align all parameters of the two histograms to make them completely comparable.

Describe the benefits and problems of such an operation.

Does the size of the plot window matter?

7.4. Dynamic Graphics

Download and install a copy of ggobi. Load the *Italian Olive Oils* data from case study H (it is packaged with the installation) and enter the 8 fatty acids into a Grand Tour. Can you separate the three areas?

How can Projection Pursuit help?

8

Missing Values

There are many reasons for the existence of missing values: the failure of a sensor, different recording standards for different parts of a sample, or structural differences of the objects observed that make it impossible to record all attributes for all observed instances. A wind sensor might stop recording values after it was damaged in a thunderstorm, different hospitals might record different properties of a patient's history, a survey on cars won't be able to state the number of cylinders for a car with a rotary engine. Figure 8.1 shows missing value plots for the *Augsburg* dataset from case study G. In a missing value plot, a bar is drawn for each variable that has missing values. The left part of the bar represents the proportion of observed cases, the right part of the bar shows the proportion of missing values for that variable. As for all area-based plots, bars representing counts of missing values are plotted in white. The left plot of Figure 8.1 shows the initial setting, listing all variables in the order they appear in the dataset. The right plot of Figure 8.1 is sorted according to

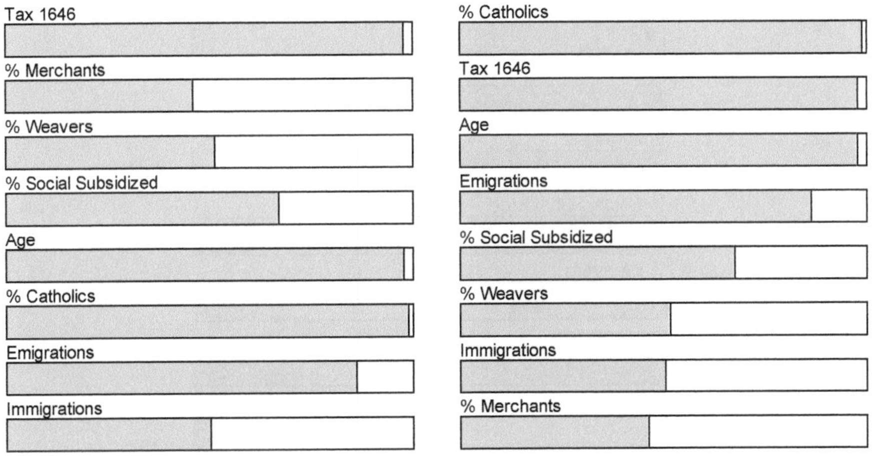

FIGURE 8.1
Missing value plots for the *Augsburg* data. Left: Initial variable order, right: variables ordered according to the number of missing values per variable.

the number of missing values.

This chapter will show how graphical methods can be used to deal with missing values and investigate their multivariate structure. Most classical statistical procedures cannot genuinely handle missing values. Thus, most of the literature on missing values talks about methods to either delete missing values from a dataset in various variations, or impute values, i.e., replace missing values with sensible estimates. Although imputation seems to be an attractive approach, one has to keep in mind that imputation methods cannot be chosen independently from a possible analysis method, as the way values are imputed will systematically influence estimates and inferences drawn from an analysis.

Monotone Missingness

Monotone missingness can be easily detected by using a single missing value plot. Monotone missingness means that for at least two variables x_j and x_k the presence of a missing value for case i in variable j implies a missing value for case i in variable k. More formally, $\forall i :$ $miss(x_{ij}) = true, miss(x_{ij}) \Rightarrow miss(x_{ik})$. Thus $|\{i : miss(x_{ij}) = true\}| \leq$ $|\{i : miss(x_{ik}) = true\}|$ holds true as well. Often a whole sequence x_{j_1} to x_{j_n} of variables can be found such that monotone missingness can be found for each pair x_{j_l} and x_{j_m} for all $l < m$. Monotone missingness

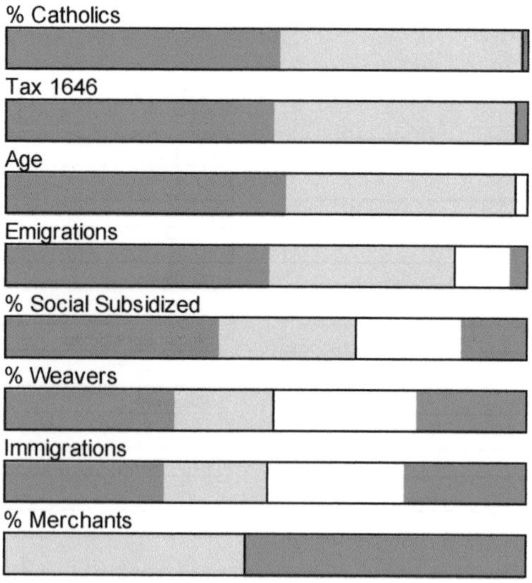

FIGURE 8.2

Missing value plots for the *Augsburg* data. To detect monotone missingness, the variables need to be sorted according to the number of missing values.

can be explored in a missing value plot by first sorting the plot according to the number of missing values and then selecting the largest to the smallest bin of missing values step by step. The property of monotone missingness holds for the selected variable and all variables with bins of the missing values completely selected.

Figure 8.2 shows the same missing value plot as shown in Figure 8.1 right. The missing values of the variable *% Merchants* are selected. From this plot we see that there is monotone missingness between the variables *Tax 1646*, *% Catholics* and *% Merchants*, i.e., that missingness in either of the two variables implies missingness in *% Merchants*. As we only look at only one, respectively two cases here, this is most probably more a coincidence than a structural feature.

Structure of Missing Values

Monotone missingness is just one special structural feature of missing values. A more fundamental question is to look for structures in missing values in general.

Three structural properties of missing values are generally discussed:

- **Missing Completely at Random (MCAR)**
  Missing completely at random describes the situation where the missing values are randomly distributed across all cases in the dataset. This setting is very unlikely in practical applications and mostly discussed in the context of imputation methods. It can be seen very easily in a missing value plot. In the case of missing completely at random, all variables must appear in the missing value plot with the same probability and have approximately the same number of missing values each. Furthermore, selecting any missing value bar in the plot (or any other subgroup of the data) should result in equal proportions of highlighting in observed and missing values for each variable.

- **Missing at Random (MAR)**
  Missing at random relaxes the generality of missing completely at random in that missingness might be different for certain subgroups of the data. However, once we control for these subgroups the MCAR property holds true within the subgroup. MAR is usually investigated between pairs of variables of which at least one must have a substantial number of missing values. Whenever there is a significant interaction between the presence of missing values and one of the variables, these missing values cannot be regarded as missing at random with respect to that variable.

  In an analysis we usually want the most important independent

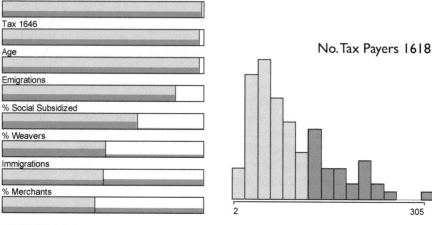

FIGURE 8.3
Missing at random does not hold true for *Emigrations, % Social Subsidized* and *% Weavers* with respect to *No. Tax Payers 1618*.

variables to be missing at random with respect to the response variables. Figure 8.3 shows all variables with missing values of the Augsburg dataset in a missing value plot. In contrast to Figure 8.2, the highlighting in this plot is rotated by 90°. All cases/districts with more than 120 tax payers in 1618 are selected in the linked histogram of Figure 8.3. The highlighting in the missing value plot shows a clear interaction for the three variables *Emigrations, % Social Subsidized* and *% Weavers*, which have fewer missing values for higher values of *No. Tax Payers 1618*.

- **Non-Ignorable Missingness**
  If neither MCAR nor MAR holds true, one speaks of non-ignorable missingness. This is the most common case for real data without trivial missing structure.

Selecting and linking among the bars of a missing values plot as well as to other plots of the data can help reveal the structure of missing values in a dataset. In practical applications, the structure of missing values can hardly be captured by the aforementioned concepts alone.

In general, most graphics can deal with the presence of missing values without any further modifications. However, the user must be aware of the missingness if no further indicators are used in the graphics.

Possible modifications of graphics are illustrated in the next section and are discussed in Unwin et al. (1996) especially in the light of interactive graphical applications. The concepts discussed in Unwin et al. (1996) are implemented in the software MANET.

Handling of Missing Values in Standard Plots

In principle, there are three ways to handle missing data in statistical graphics:

- **Assigning missing values to a specified value**
  This solution results in a plot dependent modification of the data and is just a very simple form of imputation. The idea is to plot the missing values at a specific point of the scale, where they stick out as a specific group that is clearly not part of the "normal" data.

  Figures 8.5 to 8.7 show that the constant value assigned to missing values may vary from plot to plot and within a plot's parametrization.

  The clear advantage of such an approach is that the way of plotting data remains unchanged. On the other hand, the fact that missing data are treated and plotted as regular data in the dataset and thus extend the range of the data might be misleading and result in erroneous interpretations!

- **Tolerate missing values**
  To tolerate missing values in plots by not plotting such cases sounds quite dangerous in the first place, but is far more efficient than coding missing values with pseudo values that are then included in the plots. Prerequisite of such an approach is to first investigate the missing values in a missing value plot and link this plot to all other plots during an analysis. Thus the presence of missing values remains apparent and aids the interpretation.

- **Extend statistical graphics to represent missing values**
  The most complete solution for handling missing values in graphics is to modify all graphics such that they not only tolerate missing values (as in the latter approach), but add an explicit graphical representation of the missing values.

Fortunately, the handling of missing values is trivial for all data on a categorical scale because the attribute of missingness can be simply added as an extra category labeled "NA." Figure 8.4 shows how missing values can be added as an extra category in a barchart. The variable

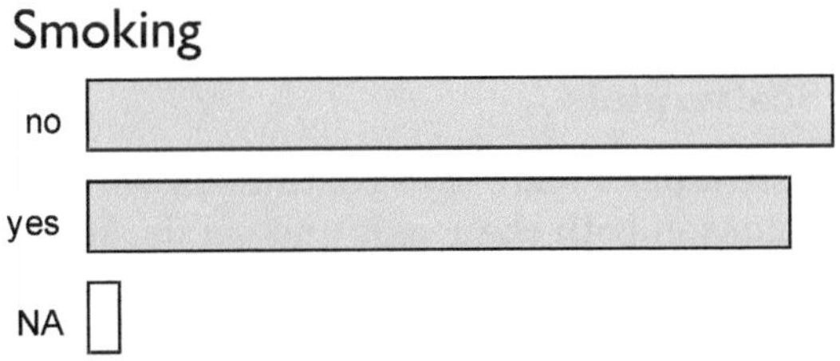

FIGURE 8.4

A barchart for the variable *Smoking* for the birthweight dataset from case study C. The missing values simply form an extra class.

Smoking from the birthweight dataset from case study C gets an extra class "NA" which is also plotted in white in order to visually discriminate it from the regular classes *yes* and *no*.

The handling for data on a continuous scale can be done according to one of the three solutions mentioned above. The different approaches are shown for different plots.

Histograms

Although histograms look quite similar to barcharts, they show continuous data and thus need different handling of missing data than barcharts. Figure 8.5 shows three versions of the histogram for the variable *Emigra-*

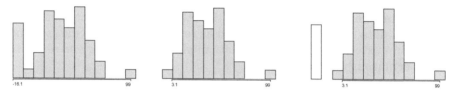

FIGURE 8.5

Three versions of plotting a histogram for a variable containing missing values. Left: missing values are assigned the minimum value −20% of the range, middle: only observed values are plotted, right: an extra bar is added to the left, showing the count for the missing values (MANET).

tions which contains 13 missing values. In the plot on the left, all missing values are assigned the minimum value −20% of the range of the variable. Although this is quite far away from the rest of the data, no gap is visible to the rest of the data, and thus will lead to a misinterpretation. In the middle plot, only observed values are plotted. The right plot shows an extra bar for the missing values which is added to the left of the histogram, outside the scale of the variable.

Scatterplots

Scatterplots show two continuous variables and thus can have missing values in both the x-axis and y-axis. In Figure 8.6 three scatterplots are shown, each using a different approach to handling missing values. The left plot replaces missing values with the minimum values minus 20% of the range. As found already in Figure 8.5, this solution can be very misleading if the analyst is not fully aware of the coding mechanism for missing values. The middle plot in Figure 8.6 only plots the observed values. The right plot shows missing values as projections onto the axes where at least one value has been observed. If both values are missing, the values

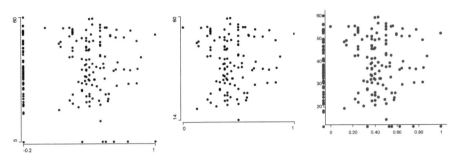

FIGURE 8.6
Three ways to handle missing values in a scatterplot. Left: missing values are assigned the minimum value -20% of the range, middle: only observed values are plotted, right: the missing values are plotted as projections on the axis where at least one of the values was observed (KLIMT).

are all plotted on the same point in the lower left corner. Without too many missing values that result in strong overplotting, this solution is the most favorable.

Boxplots

Boxplots are neither area-based plots (such as barcharts or histogram), nor do they plot a single glyph per observation. This makes boxplots special in the way they can incorporate missing values. Figure 8.7 shows four different ways missing values could be handled in boxplots. In Figure 8.7 a) missing values have been imputed as minimum -20% of the range. Because the imputed values are then entered in the construction of the boxplot (i.e., calculation of median and hinges), the boxplot is false, and the missing values can no longer be identified. Figure 8.7 b shows the boxplot that ignores missing values. In Figure 8.7 c and d the missing values are also added at minimum -20% of the range, but the boxplot is constructed without taking these values into account. Depending on whether a common scale (c) or an individual scale (d) is chosen, the extra points for the missing values align at the same level or not. To make the representation of the missing values different from ordinary outliers in a boxplot, the points are plotted in a distinctive color.

In summary, all attempts to incorporate missing values in a boxplot are very unsatisfactory, and linking a missing value plot to a standard boxplot like the one in Figure 8.7 b seems to be the simplest solution, which is least confusing and error prone.

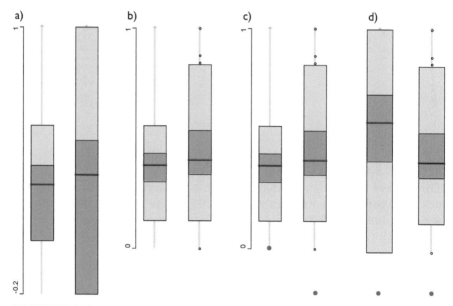

FIGURE 8.7

Four different ways of incorporating missing values in boxplots. a) assigning missing values the minimum value −20% of the range results in false boxplots; b) ignoring missing values in the plot; c) and d) missing values are placed at the minimum value −20% of the range, but do not change the boxplot (common scale c), individual scale d)).

Parallel Coordinate Plots

Parallel coordinate plots are built up by a polyline for each observation. Thus, the presence of missing values will disturb the visual pattern, which make parallel coordinates so powerful. There are two ways of dealing with missing values in parallel coordinate plots:

- Missing values are placed outside the range of the observed values and all polylines are still connected. This solution can produce many crossing lines and clutter the display, but shows a fully connected polyline for all cases.

- The polyline is broken whenever a missing value is encountered at one of its nodes. This solution will reduce clutter in the display, but a single observation may now contribute several non-contiguous polylines. For cases where the values of both adjacent axes are missing, a single glyph (dot or short line) must be plotted to represent the value.

Figure 8.8 gives an example of the latter solution for the Tour de France data from case study F. The parallel coordinate plot shows the cumu-

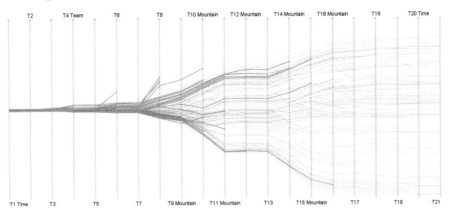

FIGURE 8.8
Missing value plots for the *Tour de France* data. Whenever a rider drops out of the classification, the line ends at that particular stage. All riders who did not complete the Tour are highlighted, and unhighlighted cases are made less prominent by using α-transparency.

lative times of all riders on the same scale, but shifted to be aligned at individual means. All cyclists who did not complete the tour — and thus have missing values for all succeeding stages after dropping out — are highlighted. In this plot, it is very easy to see at what stage and rank the cyclists had to quit the race. Assigning a constant to these missing values would disturb the clear pattern of Figure 8.8. Obviously, a situation where there is an order for the axes of a parallel coordinate plot and no missing values in between is ideal. If the polylines are broken at very many points, the readability of the parallel coordinate plot may suffer, no matter which of the above options is used to handle missing values.

Exercises

8.1. Missing Value Plot
Take the tips dataset from case study i and artificially introduce about 30% missing values.

(a) Completely at random

(b) Randomly when controlled for *Gender*

(c) Depending on the *Tiprate*; more precisely, the smaller the *Tiprate*, the more likely missing values should get.

Use the missing value plot and other linked graphics of the data to investigate the structure of the three scenarios (a) to (c). Can the graphics reveal the structure of the missing values?

8.2. Missing values in Parallel Coordinate Plots
For the *Probability Theory Exam* data from case study A, create a missing values plot, a parallel coordinate plot and histograms for the achieved points for all eight questions of the exam (students could choose 5 out of 8 possible questions, thus on average the points for 3 questions are missing).

(a) Which questions or combination of questions were most popular?

(b) Is there an order of the 8 questions in the parallel coordinate plot that minimizes the number of gaps in the polylines?

(c) Which questions were the most successful?

8.3. Imputation
Discuss the advantages and disadvantages of the different alternatives (randomly, constants, means or medians, linear models, ...) to impute missing values.

Impute the missing values for the *Augsburg* dataset from case study G. Investigate the result of the imputation with interactive graphics and describe how satisfactory the imputations are.
(To reduce the effort pick only selected variables and approaches.)

9

Large Data

The notion of a "large" dataset is quite relative. For a medical study 1,000 patients can be quite large; a typical telecommunication application won't call a problem "large" unless more than 1,000,000 cases are involved. In general, the transition from "normal" to "large" will usually take place, whenever classical tools and procedures no longer work properly. Large can mean a large number of observations, a large number of variables or both.

Many concepts in classical mathematical statistics rely on testing the significance of hypotheses. For example, a χ^2-test of independence can always be rejected if only the sample is large enough. Thus for large datasets, mathematical statistical applications tend to be significant no matter how small the effect is. A distinction between statistical significance and practical relevance becomes important in these cases. More modern techniques such as cluster analysis algorithms have a complexity of at least $\mathcal{O}(n^2)$ and thus will hit computational limits sooner rather than later.

This chapter investigates the problems of statistical graphics when dealing with large datasets and looks at possible modifications.

9.1 Unaffected, Summary-Based Plots

All area-based plots such as barcharts, mosaic plots and histograms display a summary of the variables. Thus their complexity usually does not change with the number of observations which are summarized. For example a barchart for *Gender* can be displayed for a school class of 20 as well as for the complete U.S. Census; the number of categories will always be two. We can distinguish three different groups of variables according to the way the number of categories scales along with the number of observations.

1. The first group summarizes all variables which are known to have a fixed number of categories such as *Gender* or *US State*. This number will not change no matter how many entities we summarize and a

sample will usually cover all categories.

2. The second group includes all variables where we expect an increasing number of categories with an increasing number of cases, however, with an a priori known upper limit that will not be exceeded. This group includes variables such as *Make of Car* or *Native Country*.

3. The last group includes all variables where the only reasonable theoretical limit to the number of observed classes is given by the number of cases. In this situation, the number of observed classes is usually close to the number of recorded cases. Typical examples for this kind of variable could be *Favorite Football Team* or *Native City*.

based on 63,756 observations   based on 510 observations

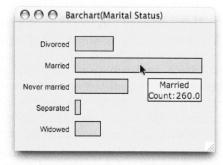

FIGURE 9.1

Two barcharts from the *Current Population Survey '95*. Left: the complete sample of almost 65,000 observations, right: the 510 samples taken in Delaware.

Figure 9.1 shows two almost identical barcharts for *Marital Status* of the *Current Population Survey '95*. The left barchart shows the chart for the complete sample of almost 65,000 observations, whereas the right chart only shows the 510 samples taken in Delaware. The queries reveal a factor of more than 100 in scale of the two plots. Whereas the barchart is not affected when the sample size scales up, the interpretation of the variability of the bars has to change. Scaling problems arise for barcharts and even more so for mosaic plots, whenever the number of categories to plot becomes large. In this case, interactive support for sorting barcharts (see Section 7.1) and transformations of categorical data (see page 42) are essential to manage large datasets. Sorting can help for barcharts, but not for mosaic plots. For an example of logical zooming in mosaic plots see Unwin et al. (2006).

9.2 Glyph-Based Plots

In contrast to area-based plots, graphics using a single glyph for each observation have a complexity which scales at least linearly in n, the number of cases observed.

Boxplot

The core of a boxplot — the boxes and the whiskers — is based on robust statistics in the form of the median, the upper and lower hinge. The difference between hinges and quartiles can be neglected for large data, such that we will work with quartiles from now on. All these statistics are invariant with respect to large n. Not so the definition of the outliers. Looking at a standard normal distribution, the inter-quartile-range, i.e., $x_{0.75} - x_{0.25}$ is 1.348. The whiskers thus extend to -2.698 and $+2.698$. All

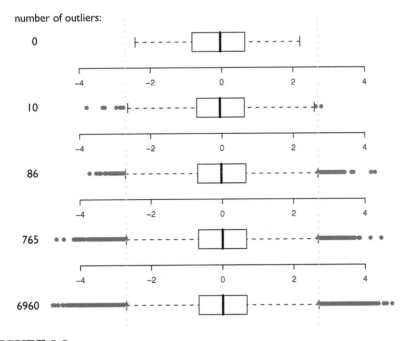

FIGURE 9.2
Five boxplots for samples from the standard normal distribution, with 100 (top), 1,000, 10,000, 100,000 and 1,000,000 (bottom) observations. The theoretical limits for outliers are marked in orange.

values outside this interval — which is 0.69% of the sample for a standard normal distribution — will be marked (and plotted) as either outlier or even far outlier. Thus the number of outliers in a boxplot grows linearly with the number of cases plotted in the boxplot. Figure 9.2 shows an example of standard normal samples with growing sample sizes from 100 to 1, 000, 000. Whereas the boxplot for 1, 000 cases — with its 10 outliers — shows no artifacts, the boxplot for the sample of 10, 000 cases shows the typical extension of the whiskers by outliers. Nevertheless, we would call the two values bigger than 4 in the middle boxplot outliers in the classical sense. The situation gets worse as we plot more points. The bottom sample of 1, 000, 000 points does not even show any outliers in the typical sense (very low density in the neighborhood and a clearly visible gap to the points toward the median). Obviously, it makes sense to think of modifications of the classical boxplot definition (cf. Exercise 9.2) or at least one needs to be aware of this phenomenon.

Parallel Coordinate Plot

Parallel coordinate plots draw their information along one dimension and thus are very similar to boxplots. Connecting the coordinate axes via lines adds the plot-specific information, but also suffers badly from over-plotting even for a few thousand cases.

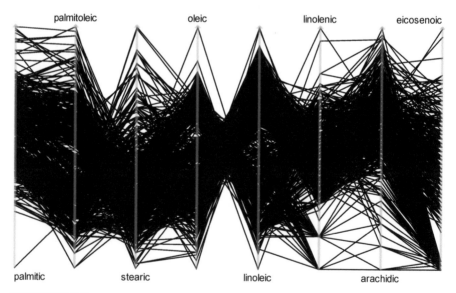

FIGURE 9.3

With fully saturated lines, a parallel coordinate plot of just about 600 cases masks most of its internal structure by overplotting.

The severity of overplotting can be illustrated with the olive oil data from case study H, which consists of only 572 cases. In Figure 9.3 all lines have been plotted in fully saturated black, masking all structural information within the black band in which most of the cases fall. The same data can be be found in Figures 4.6 to 4.9. These figures all use α-transparency (see Section 2.2 for more details on α-transparency) to make the plot more readable and show the group-structure of the data.

Scatterplot

Although scatterplots use two dimensions to lay out the data, overplotting issues cannot be avoided in scatterplots, either. Whereas the use of α-transparency as shown in Figure 3.18 was aimed toward a most efficient estimation and display of the 2-dimensional density of the scatterplot, using α-transparency for large data might be the only way to effectively display the data. Figure 9.4 shows four scatterplots of chess rating data, i.e., *Rating* vs. *Year of Birth*. The upper left scatterplot illustrates the naïve default settings of a relatively large dot and no transparency. Clearly, this representation does not give any insight into the data. The upper right and the lower left scatterplots use an α-transparency of 0.16 and 0.01, respectively (note that the α-values are only a rough guide for comparison, because the visual effect of α-transparency may vary between output devices, e.g., the computer screen and different printers). Reducing the α-value reveals two thresholds in the data. The number of cases increases rapidly for birthdates in the early 1950s; the lack of cases in the early 1940s is only barely notable. There are only a few cases with a rating below 2,000, which drops to a rating of 1.700 for birthdates starting in 1985 and later. Both features are invisible in the saturated scatterplot.

The lower right scatterplot shows an alternative to α-transparency, a so-called binned scatterplot. A binned scatterplot is in principle nothing other than a histogram in two dimensions. In a binned scatterplot, the plotting region is divided by a regular grid, and the number of observations that fall into a specific bin is counted for all bins. A gray-value is assigned to each bin depending on the number of cases observed in the bin. In Figure 9.4 lower right, darker grays correspond to larger bin counts, in order to match the representation of the other three scatterplots. The great advantage of binning over the use of α-transparency is the fact that the graphical complexity of the plot is only determined by the number of bins in the scatterplot and is no longer a function of the number of observations displayed. On the other hand, a binned scatterplot is a completely different plot than the traditional scatterplot. For binned scatterplots (cf. Carr et al. (1987) for an early reference) the shape of the bins and the mapping of counts to gray shades have to be specified. A drawback of binned scatterplots is aliasing effects, which occur whenever the resolution of the data does not match the binning grid —

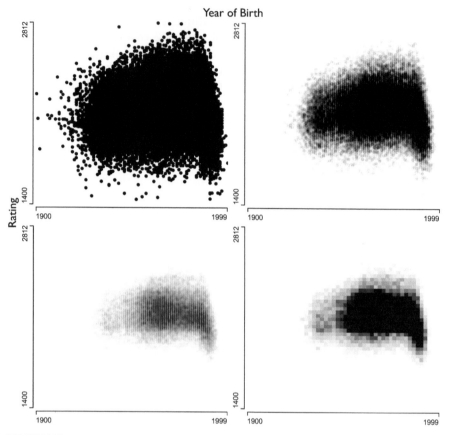

FIGURE 9.4
Different levels of α-transparency (1.00, 0.16 and 0.01) and binning for
69,541 chess ratings (top left to bottom right).

e.g., when age is measured in years from 1 to 90 and this information is
binned over 50 bins, most bins will cover 2 years and a few will cover only
one year.

   Many of the benefits of binned scatterplots for large data can also be
obtained from traditional scatterplots by introducing α-transparency and
variable point-sizes. In particular, the scatterplots in the lower row of
Figure 9.4 show almost identical features. Regardless of the plotting technique chosen the most important issue is the possibility to interactively
change the plot parameters in order to obtain the most revealing view of
the data.

Exercises

9.1. Significance and Relevance in Barcharts

Look up the current proportions of males and females in the U.S. at an official Census site (cf. Exercise 6.3), and construct a simple test for $H_0 : p = 0.5$ vs. $H_1 : p \neq 0.5$.

For what lowest n does your test reject the null-hypothesis given an α-level of 0.05?

Discuss the consequences of the above results for the interpretation of large samples.

9.2. Scaling up Boxplots

   (a) For a log-normal distribution, calculate the expected number of outliers of a sample of size 100,000 on both sides of the distribution.

   (b) Discuss possible modifications for the outlier definition of boxplots for large data based on the

   i. number of outliers

   ii. distance to the adjacent points

   iii. density.

9.3. Searching the Haystack?

Open the *Pollen* dataset which can be obtained from the book website. The *Pollen* data was artificially created for the 1986 ASA Data Exposition by David Coleman of RCA Labs.

Try to find the "special feature" in the dataset by using first parallel coordinate plots and second scatterplots utilizing α-transparency, zooming and point-size variation.

10

On the Examples

Initially, five out of the nine examples were planned to come from Cox and Snell (1981). When working out the case studies only the *Detergent* data remained; the other four datasets were abandoned in favor of more modern datasets with more interesting features.

Two datasets contain purely categorical variables, two purely continuous variables, and five datasets have variables on mixed scales. Three of the examples have a geographical reference.

Each of the nine examples covers eight to ten pages, with usually half of them being graphics of the most important steps and findings in the graphical analysis. For each dataset, the following six points are discussed:

1. **Background**
   What are the most important facts about the background? What information do we need to understand the variables and the study goal?

2. **Goals of Study**
   Why has the data been collected, or what are the main questions we might want to answer by an analysis?

3. **Description of Data**
   What is the source of the data, and what are the variables about? Which data preparation does it take to eliminate errors and/or derive important information?

4. **Graphical Analysis**
   Which plots and interactions does it take to analyze the most important issues of data graphically?

5. **Further Analysis**
   How can further analyses — often of parametric nature — give additional insight and complement the graphical analysis? Are there other statistical methods and algorithms which deal with the matter more efficiently? Can we get further or even different findings there?

6. **Result**
   What were the key findings of the graphical analysis and could the
   initial questions be answered? Did new questions arise? Was the
   graphical analysis sufficient? Which issues do we want to verify
   and/or quantify with parametric methods? Are there further issues
   we might want to look into, more data we need to collect?

Additionally, there are Exercises for each case study which cover points
not discussed in the case study so far or discuss issues that are worthwhile to look at in more detail.

Although the graphics in the examples are not screen dumps from the
actual analysis, they were all generated from actual plots used during an
analysis. Thus each of the graphics can be generated in the exact same
form using the Mondrian software.
   A valid question to raise is how can an interactive analysis be published in a static medium like a book? The answer is twofold. We are

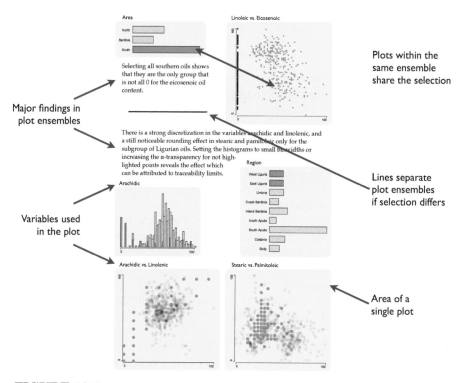

FIGURE 10.1
A sample plot ensemble as used in the examples taken from the Olive Oil
case study H.

accustomed to learning rather interactive techniques from books. Every book on cooking does so. The important point is the fact that we are actually cooking or, in our case, analyzing data. In doing so, we learn the necessary steps to achieve satisfying results, finally without following a specific recipe or a specific example analysis.

Figure 10.1 shows an example of the graphical conventions used in the case study part of the book.

All graphics pages in the case studies follow these conventions: Plot ensembles are separated by lines (if the selection differs from the remaining plots) and contain all plots that contribute to this ensemble. If highlighting information is present in the plots, all plots within an ensemble share the same highlighting. Using the highlighting information in the plots underlines the interactive nature of the analyses. Since most plots are not labeled as presentation graphics would be, the variables in the different plots are annotated in the plot title. For scatterplots the common notation "y vs. x" is used, i.e., the variable on the y axis is named first, the one on the x axis second. Furthermore, a small paragraph of explanation summarizes the most important finding of the plot ensemble.

In times where "data mining" and "knowledge discovery in databases" are still buzz-words, the size and complexity of most of the case studies seems to be rather small. Most of the datasets have only hundreds to some thousands of instances and no more than dozens of variables. Although this may sound small, it is still big when compared to the typical examples in a textbook on statistics. Furthermore, most of the interactive graphical methods presented here scale quite easily to much larger datasets. Existing scaling issues were discussed in Chapter 9 of the principles part.

Although we recommend working through the principles part first — or at least working through both parts in parallel and starting with easier case studies — more experienced data analysts with a solid knowledge of statistical graphics may enjoy reading the case studies and only refering back to the principles part when needed.

Part II

Examples

A

How to Pass an Exam

Background

Exam testing is a common means to assess students' knowledge and skills. Not only do different educational systems have different ways to hold and grade exams, standards change over time as well. With few students to test, oral exams are clearly the least  intricate and least time consuming way to grade and rank students. With hundreds of students in a course — unknown in the U.S., but quite common in Western Europe during early semesters — written exams are the only solution.

Multiple choice tests consist of a list of possibly correct answers for each question and the student chooses (marks) one (or more) answers each. Such tests are increasingly attractive with more students to test as grading becomes a matter of comparing the choices made by a student to the 'true' set of answers. Multiple choice questions can cover a far broader range of topics than fewer free-response questions. On the other hand, multiple choice tests may overemphasize memorization and test processes and comprehension poorly. They usually leave no room for disagreement or alternate interpretation, which makes them particularly unsuitable for non-technical subjects.

Goals of Study

Given that most university exams are not defined by a central instance, it is worthwhile to check how well an exam captured the knowledge of the students. In particular, has the preparation for the exam by homework worked out well? Are certain questions too complex or too easy, or can the structure of the exam — multiple choice vs. free-response — be optimized? Depending on the results, the exam should be changed to better assess the student's skills.

Description of Data

The data we look at in this case study reflects the results of a 3-hour written exam in probability theory, which is taken mainly by math students in their third semester of studies. The exam was conducted in the winter semester 2005/06 at Augsburg University. In addition to the lecture, students were prepared in 4 homework groups. The exam consists of a multiple choice part (weighted 1/6th) and 5 out of 8 questions (weighted 5/6th) free for the student to choose. For 62 students who took the test, the following data was recorded:

- **Gender**
- **Major Subject**
  Mainly Math or Business Math
- **Semester**
  Students should take the test in the third semester, but can also take the test in a later semester, especially if they failed their first attempt or want to improve their mark.
- **Homework Group**
  One out of four; student's choice according to their class schedule.
- **Pre-Score**
  The average mark on the 12 assignments ranged between 0 and 25 points.
- **Multiple Choice**
  Points achieved in the multiple choice part of the exam, 0 – 50.
- **Points in Question 1 – 8**
  0 – 50 points, missing, if question was not selected.
- **Question 1 – 8 Selected?**
  Binary variables (Yes/No) indicating whether a question was chosen or not.
- **Total Selected**
  Number of questions worked on.
- **Sum Points**
  Sum of the multiple choice part and the 5 best questions worked on.
- **Mark**
  According to the German educational system, ranging between 1 (best) and 5 (failed) with ±0.3 differentiations.
  (There is no "0.7" or "5.3" and any score lower than 4.0 is a failing grade.)

Graphical Analysis

This analysis will only cover some of the most interesting of the numerous structural features of the dataset.

We start with the most important univariate features, which describe the sample structure. The barchart for *Gender* shows that about 1/3 of all students are female. Most of them (71.4%) study Business Math, which is the most popular subject with more than half of all students. Another quarter of all students registered for Math. The barchart for *Group* shows that the four homework groups differ strongly in the proportion of female students. Whereas group 1 has more than 50% female students, group 3 (the smallest group) has less than 20%. The fluctuation diagram for *Subject* and *Group* shows that Business Math students can be found primarily in groups 1 and 3 and Math students in groups 2 and 4. Such a structure is to be expected, because different subjects imply different class schedules, such that some of the homework groups overlap with other courses. The histogram of *Pre-Score* shows three groups of students: (1) those who regularly handed in their homework, (2) those who only rarely worked on the homework and (3) those who almost never handed in their homework. The corresponding spinogram reveals no relevant structure, given the small sample size. Only half of all students actually took the test in the scheduled semester as indicated in the barchart for *Semester*.

The missing value plot immediately reveals the popularity of the 8 questions. Questions 5 and 6 are by far the least popular choices, whereas almost every student worked on question 7. The parallel boxplots for the results of the individual questions show median results between 27 and 30 points for questions 2, 4, 6, 7 and 8. The three remaining questions have median results of 20 (question 3), 5 (question 1) and 0 points (question 5), which matches quite well with the insights from the missing value plot. Both distributions of the multiple choice part as well as of the total points achieved are left skewed. Comparing the overall results between the groups in a boxplot of *Total Points* by *Group* shows the best results for group 2 and the worst results for groups 1 and 4.

We use a scatterplot to test how well *Pre-Score* can predict the total points in the exam. Adding a linear regression gives a relatively poor R^2 of 20%. This is mainly due to the three groups that can be found in the distribution of the pre-score. Selecting group 2 shows a far stronger association for this group with an R^2 of 73.9%. The number of semesters has a surprisingly strong effect on the overall result. The boxplot *Total Points* by *Semester* shows a decline of roughly 30 points per year and a penalty of almost 50 points for students who started in the summer term and not in the winter term which is the default. To check whether the multiple choice part of the exam adds a discrimination of the result which is not already captured by the 5 questions, we derive the variable *Total Points - Multiple Choice*.

Gender

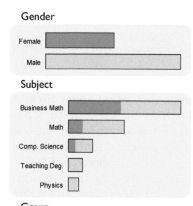

Subject

About 1/3 of the students are female. Business Math makes up for 50% of all students and has an almost equal share of male and female students. Only a few students come from other subjects.

Subject

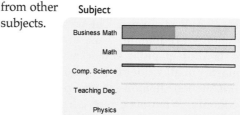

Group

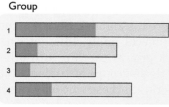

Group 1 is the largest of all four groups. Groups 1 and 4 have a higher rate of female students. Group 1 is dominated by female Business Math students. Physicists can only be found in Groups 3 and 4.

Subject x Group

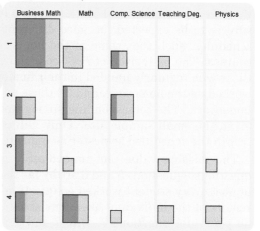

Pre-Score

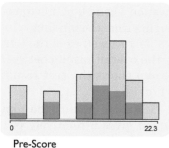

There are three groups in the distribution of the Pre-Score. The majority worked on their homework regularly and reached an average score of 15 points on average. Others handed in only a few assignments ending up with no more than 6 points. A third group almost never got any points on their homeworks.

Pre-Score

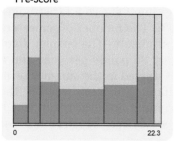

50% of all students took the exam in the 3rd semester as planned.

Semester

Missing Values

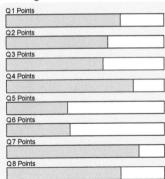

The missing value plot shows the popularity of the questions. Q7 was most often selected, whereas Q5 was least popular.

Multiple Choice

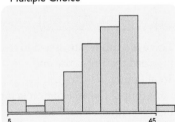

The distribution of the multiple choice part is left skewed with a mode at 35 points.

The distributions of the eight questions can be compared in parallel boxplots.

Q1 and Q5 show very poor results, whereas the other six questions have median results between 20 and 30 points.

None of the students got the full number of points for Q1, Q5 and Q6.

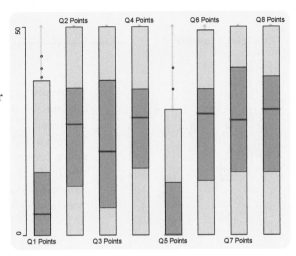

Total Points | Group

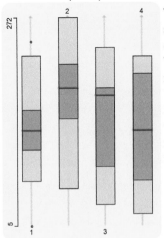

The four groups show quite different results. The median result of groups 1 and 4 is about 50 points below the median result in groups 2 and 3.

Total Points

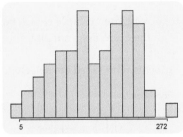

The distribution of Total Points is also left skewed with a mode at 200 points but has an extra mode at 120 points, which corresponds to "barely passed."

Group

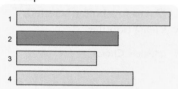

Total Points vs. Pre-Score

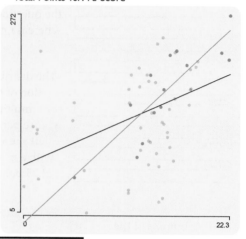

There is a moderate association between the Pre-Score and the Total Points (R^2=20%).
For group 2 the association is far tighter with an R^2 of 73.9%.

Total Points | Semester

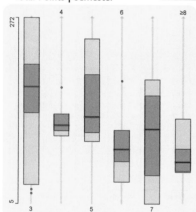

The results depend strongly on the semester in which the test was taken.
There is a decline by roughly 30 points per year, and students who took the test in an even semester performed worse by a margin of about 50 points.

Removing the multiple choice part from the exam does not alter the result.
Students who failed are selected.

Total Points vs. (Total Points - Multiple Choice)

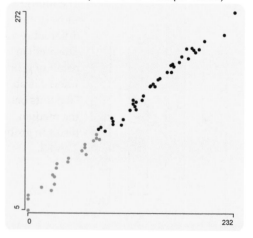

Total Points

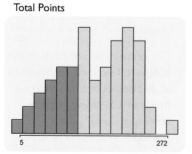

In the scatterplot of *Total Points* vs. *Total Points - Multiple Choice* we see a very strong correlation; which is to be expected. Only one student who failed would rank before a student who passed, when using the result without the multiple choice points.

Further Analysis

In the graphical analysis, we found that students in later semesters and those who started in summer term have on average inferior results compared to students in their third semester. We set up a simple linear model in R (assuming the data to be in the dataframe called PTE) to estimate the effect, which we already read roughly from the boxplots.

```
# Make all >=8 to 8 and the factor numeric
#
> levels(PTE$Semester)[6] <- "8"
> PTE$Semester <- as.numeric(as.character(PTE$Semester))

# Dummy for even years
#
> even <- 1 - (PTE$Semester %% 2)

> ll <- lm(Total.Points ~ Semester + even, data=PTE)
> summary(ll)
...
Coefficients:
            Estimate Std. Error t value Pr(>|t|)
(Intercept)  200.273     21.927   9.133 6.85e-13 ***
Semester     -12.623      5.063  -2.493   0.0155 *
even         -23.909     21.633  -1.105   0.2736
...
> plot(PTE$Semester, PTE$Total.Points)
> abline(ll$coeff[1:2])
> abline(ll$coeff[1]+ll$coeff[3], ll$coeff[2])
```

The model estimates a decline of 12.6 points per semester, i.e., about 25 points per year. Students who are in an even semester — who started in summer term — have an estimated extra penalty of 23.9 points. These estimates are slightly smaller than what we read from the medians of the boxplots, which might be explained by the left-skewed distribution.

Summary

Graphical methods are well suited to explore the structure of the sample. Using boxplots *y* by *x* and parallel boxplots the result of the exam can be conditioned on the various factors. Furthermore, it turned out that the

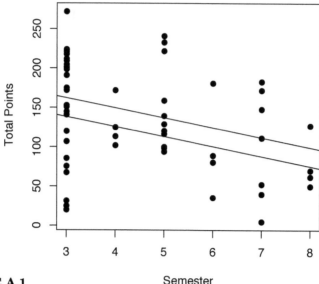

FIGURE A.1

The linear model estimate.

pre-score is a good indicator for the exam result for those students who take the homework seriously. The multiple choice part of the exam might also be neglected, as it does not contribute much to the discrimination of the students' skills.

The examination of the popularity of certain combinations of questions is left as an exercise.

Exercises

1. Is *Pre-Score* correlated with *Semester*? Test graphically with linked barchart and histogram/boxplot and a boxplot y by x.

2. What can be said about the number of questions a student worked on? Five questions are the norm — does it help to try out more questions or focus on fewer?

3. Analyze the structure of combinations of questions using the binary variables in a mosaic plot. Which combinations are most popular; which are most successful?

4. Compare the four homework groups regarding their average result and the number of students passed (students needed at least 110 points to pass).

5. Investigate and describe the influence of *Gender* and *Subject* on the students' results. Can these factors be added to the model set up above?

B

Washing — What Makes the Difference

Background

Water plays a major role in the process of cleaning laundry. Water level, temperature and softness are the three main factors that influence the washing result. Whereas the first two can be controlled relatively easily by the operator of the washing machine, the third — water softness — is a parameter which is fixed and determined by the local water source.

In general, three temperatures are distinguished on fabric care labels. Cold or Cool refers to temperatures between 65° and 85°F (18°–30°C). Warm water ranges between 85° and 120°F (30°–50°C). All temperatures above 120°F (50°C) are regarded as hot.

Interestingly the washing culture is quite different in central Europe. The traditional distinction between cold, warm and hot is the same, but refers to the temperatures 30°C (120°F), 60°C (140°F) and 90°C (195°F). In order to save energy and to go easy on the fabrics, the use of improved detergents nowadays allows us to wash almost all fabrics at 60°C (140°F) with the same satisfactory results.

The phenomenon of different levels of water softness (or hardness) is not noted by many people as long as they deal with soft water. Only recently the scale on which the degree of water hardness is measured has been standardized as mmol/L (millimoles per liter) calcium and manganese ions or the mg/L calcium carbonate equivalent. Various obsolete (national) degrees are:

- Clark degrees (°Clark) / English degrees (°E)

(Conversion to mg/L calcium: divide by 0.175. One degree Clark corresponds to one grain of calcium carbonate in one Imperial gallon of water which is equivalent to 14.28 parts calcium carbonate in 1,000,000 parts water.)

- German degrees (°dH)
  (Conversion to mg/L calcium: divide by 0.14. One degree German corresponds to one part calcium oxide in 100,000 parts of water.)

- French degrees (°f)
  (Conversion to mg/L calcium: divide by 0.25. One degree French corresponds to one part calcium carbonate in 100,000 parts of water.)

- American degrees
  (One degree American corresponds to one part calcium carbonate in 1,000,000 parts water (1 mg/L or 1 ppm).)

There are several rules that classify ranges of water softness to fixed terms. The most common are:

- **Soft**: 0 – 40 mg/L as calcium

- **Medium:** 40 – 80 mg/L as calcium

- **Hard:** 80 – 120 mg/L as calcium

- **Very Hard:** >120 mg/L as calcium

Goals of Study

The study, which dates back to Ries and Smith (1963), compares two detergents, a new product "X" and a standard product "M." The goal was to find out under which circumstances an individual will prefer the new product "X" over the standard "M." This is a fairly common investigation whenever the acceptance among buyers must be tested upon introduction of a new product.

Description of Data

The dataset consists of 1,008 individuals who recorded 4 attributes. *M-User* is a binary variable which indicates whether an individual previously has been using product "M" or not. *Temperature* used during the test, measured at the two levels *low* and *high*. *Water Softness* was recorded at the levels *soft*, *medium* and *hard*. *Preference* indicates preference for product *X* or product *M*.

The 4 variables make up $2 \times 2 \times 3 \times 2 = 24$ different combinations of levels, and thus can be summarized in a table with 24 entries:

M-User	Temperature	Preference	Water Softness		
			hard	medium	soft
no	low	X	68	66	63
		M	42	50	53
	high	X	42	33	29
		M	30	23	27
yes	low	X	37	47	57
		M	52	55	49
	high	X	24	23	19
		M	43	47	29

Graphical Analysis

We first look at the one-dimensional distributions of the four variables using barcharts. All variables are almost uniformly distributed, except for *Temperature*. Almost two-thirds of the individuals wash at low temperature. To study the possible influence on the preference, we select the cases for preference *X*. Observing the highlighting in the three other barcharts shows the possible interactions between the influencing factors on the preference. We find almost independence between the softness of the water and the preference. The is an interaction of low temperatures with higher preference for *X* and an even stronger interaction of previous users of *M* and the preference for *M*.

The three mosaic plots for the three independent influencing factors *M-User*, *Temperature* and *Water Softness* show independence except for *Water Softness* and *Temperature*. There is a clear tendency for washing at higher temperatures for harder water.

We now investigate the relationship between *Preference* and *M-User* on page 169. The topmost mosaic plot shows the interaction between *Preference* and *M-User*. Note that the classical splitting order (x, y, x, ...) has been changed to y first then x. This makes the plot more easily comparable to the next two mosaic plots which look at the relationship between *Preference* and *M-User* conditioned on the levels of *Temperature* and *Water Softness*. When conditioned on *Temperature* it turns out that the interaction is weaker for lower temperatures. Conditioning on *Water Softness* shows that there is almost no interaction for soft water.

We already found out that there is a tendency to wash at higher temperatures with harder water. This suggests that we look at the four dimensional mosaic plot, which shows the interaction of *Preference* and *M-User* conditioned in the six levels of *Water Softness* and *Temperature*. This mosaic plot (page 170 top) shows what the two three-way plots on page 169 suggest. The joint effect of *Water Softness* and *Temperature* is even stronger than the two single effects. *Preference* and *M-User* are independent for soft water and low temperatures. As water gets harder and warmer, the interaction gets stronger and stronger. If we were to model

Preference

M-User

Temperature

Water Softness

The preference for product "X" is more likely for non M-Users and in cases where a lower temperature was used.

Water softness does not seem have to an obvious effect on the user preference.

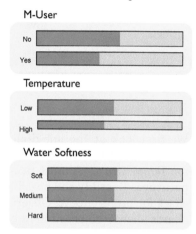

Temperature x M-User

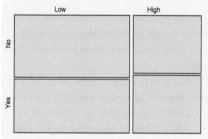

Water Softness x M-User

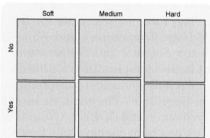

The three interactions between the three factors that potentially influence the preference.

Whereas Temperature and M-User as well as Water Softness and M-User are almost independent, there is a clear trend of washing at higher temperatures for harder water.

Water Softness x Temperature

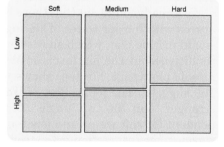

Preference x M-User

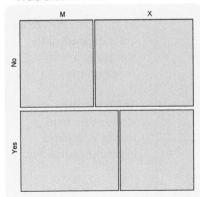

There is a relatively strong interaction between the product preference and the previously used product.

(Note: The splitting order has been changed to first y then x, to match the interaction structure with the conditioned plots. The size of the plots has been chosen to achieve comparable aspects ratios.)

Temperature x Preference x M-User

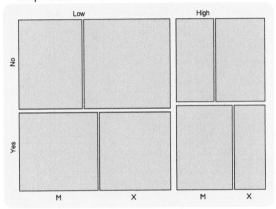

The interaction between Preference and M-User is less strong for lower temperatures.

Water Softness x Preference x M-User

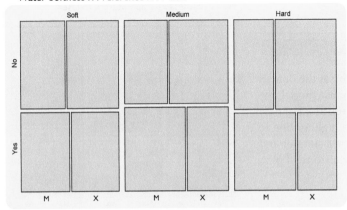

The interaction between Preference and M-User gets weaker and weaker for softer water.

Water Softness x Temperature x M-User x Preference

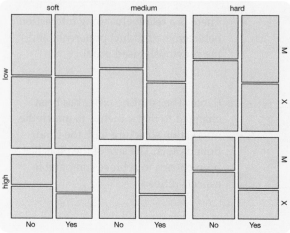

M-User and Preference are independent for soft water and low temperatures.

The interaction gets stronger for both harder water and higher temperatures.

Water Softness x Temperature x M-User x Preference

The predicted values of the model that includes the inter-action between M-User and Preference as well as Water Softness and Temperature fits the data quite well.

Although no longer significant, this model does not capture the above-described structure.

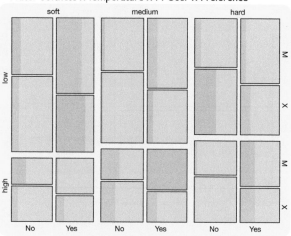

Preference

When looking at the data as a binary response model, the preference information may be linked via highlighting to achieve a simpler and easier-to-read plot.

Water Softness x Temperature x M-User

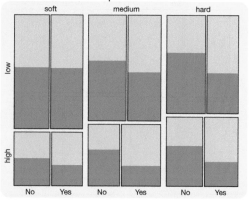

this effect, we would need a four-way interaction, which would completely specify the data, with no degrees of freedom left.

A simpler way to display the data is to put only *Water Softness*, *Temperature* and *M-User* into a mosaic plot and link on a level of the binary response *Preference* via highlighting.

Further Analysis

This dataset was re-analyzed not only in Cox and Snell (1981) pp. 86–90, but also in Fienberg (1985) pp. 71–80 and Venables and Ripley (1999) pp. 196–198. We want to look at the analysis of Venables and Ripley a bit closer.

There is actually no need to create and load a dataset that contains the 1,008 entries for each observation. Instead, the summarized dataset can be generated by

```
> deter <- cbind(expand.grid(Preference=c("X","M"),
            Temp=c("Low","High"), M.user=c("No","Yes"),
            WaterSoft=c("Hard","Medium","Soft")),
                Fr = c(68,42,42,30,37,52,24,43,
                       66,50,33,23,47,55,23,47,
                       63,53,29,27,57,49,19,29)))
> deter$WaterSoft <- ordered(deter$WaterSoft,
                    levels=c("Soft","Medium","Hard"))
```

With this dataset the log-linear model resulting from a stepwise forward procedure can be calculated by

```
> deter.ll <- glm(terms(Fr ~ M.user*Temp*WaterSoft +
                    Preference*M.user*Temp, keep=T),
                family=poisson, data=deter)

> summary(deter.ll, correlation=F)
...
Coefficients:
                        Estimate Std.Error z-value Pr(>|z|)
Preference              -0.30647  0.10942  -2.801  0.00510
M.user:Preference        0.40757  0.15961   2.554  0.01066
Temp:Preference          0.04411  0.18463   0.239  0.81119
M.user:Temp:Preference   0.44427  0.26673   1.666  0.09579
```

The output was shortened to focus on the relevant term including *Preference*. From this model Venables and Ripley conclude "*From the sign of the* M-User-*term previous users of* M *are less likely to prefer brand* X. *The interaction term, though non-significant, suggests that for* M *users this proportion differs for those who wash at low and high temperatures.*"

Summary

The graphical analysis shows a clear association between the softness of the water and the temperature people wash at: harder water needs higher temperatures to achieve equivalent washing results. The overall preference for M found by previous users of M cannot be found at "ideal" washing conditions. The harder the water gets, and the higher the washing temperatures, this preference turns out to be quite strong.

The further parametric analysis could not reveal any further insights which were not already obvious from the graphics used before.

Exercises

1. How many interactions of size 2, 3 and 4 can be displayed for the detergent data? Which of them are displayed in a mosaic plot in the graphical analysis section?
   Create all mosaic plots of dimension two and three, which are not displayed in the graphical analysis section. Do you find further insights?

2. Create a generalized linear model for *Preference* using the R-function `glm()` with a logistic link function, i.e., the `binomial` family. What is the difference in this approach from the one in the further analysis section, and how do the results differ?

3. Cox and Snell (1981) end their case study as follows: "... *however, the fact that the changes in proportion, except for the effect of previous usage, are relatively small may mean that the effects, even if real, are unimportant.*"
   Discuss the difference between statistical significance and relevance for a real effect in the light of potentially large target populations.

C

The Influence of Smoking on Birthweight

Background

It is known that many factors influence the health of a fetus. In its development during pregnancy it is supported by the mother's cardiovascular system and thus directly influenced by the mother's behavior.* Several studies have been commissioned that investigate the effects on the development of a fetus caused by substances used by mothers during pregnancy such as alcohol, drugs or tobacco.

The adverse effects of alcohol during pregnancy universally accepted today, such as the fetal alcohol syndrome, a disorder of permanent birth defects, were discovered in the late 1960s after multiple studies were performed. Similar studies were performed with focus on other substances, but also more general studies monitoring various effects on fetal development of children.

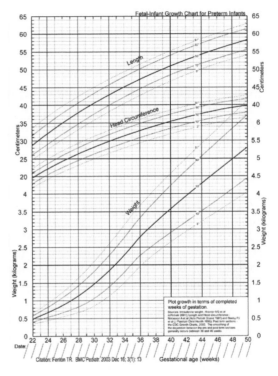

One large study was the Child Health and Development Study. It included pregnancies in the Kaiser Foundation Health Plan in Oakland, California between 1960 and 1967. During the study pregnant women were interviewed early in their pregnancy and behavioral, health-related and demographic information was collected. This study was unique in

---

*Although the placenta prevents direct contact of the embryo's blood and the mother's blood, it passes many substances unchanged.

the breadth of the information acquired and also for the fact that the interview was performed very early in the pregnancy and thus the data collection was not influenced by the (potentially unsuccessful) outcome.

The main measurement of the success of the pregnancy is the outcome (birth) and the birthweight of the newborn. Abnormal birthweight often indicates complications during pregnancy leading to an unusual rate of development. Macrosomia (big baby syndrome) may cause complications during birth and thus caesarean section is used. Nonetheless, is does not otherwise impact the infant's health in most cases. Low birthweight, however, is associated with growth restriction and some studies show a possible negative impact on the child. Low birthweight incidence is often higher in less developed countries.

Study Goals

First we want to check the quality of the data, any suspicious outliers or data points that are not realistic. Then we can look for patterns in the joint attributes such as descriptive or demographic data. Finally we will concentrate on the effects of various factors on the birthweight.

Description of Data

The data used in this example are a subset of the full study taken from-Nolan and Speed (2000). It contains a subset of a study consisting of 1,236 male, single births which survived for at least 28 days. In addition to the birthweight, several descriptive and demographic attributes were collected for both the mother and the father of the child. The full list of variables follows:

Child-related variables:

Id Number - identification number

Date - date of birth

Gestation - length of the gestation (in weeks, fractions represent days)

Birthweight - weight (in grams)

Mother-related variables:

Race - race

Age - age (in years)

Education - education (7 levels)

Height - height (in inches)

Weight - weight (in pounds)

Smoking - smoking behavior

Smoking Amount - number of cigarettes smoked

Quit - history of smoking (if quit, when)

Father-related variables:

Father Race - race

Father Age - age (in years)

Father Education - education (7 levels)

Father Height - height (in inches)

Father Weight - weight (in pounds)

Family-related variables:

Marital Status - marital status

Income - family income (in categories)

The data are a mixture of continuous and categorical variables. Some of the variables, such as weight and height of the father have missing data.

Graphical Analysis

As a first step we should look at univariate and bivariate plots to assess the quality of the data. We will notice two outliers in with unusually short gestation time (about 21 and 26 weeks). The usual gestation time is approximately 40 weeks. Although it is possible for babies to be born so prematurely, they would have to have very small weight. However, a look at the scatterplot of *Birthweight* and *Gestation* shows that they are recorded to have a normal birthweight of about 3.3 kg and 3.1 kg, respectively. This combination does not sustain the plausibility test so we have to assume that it is a data error and we will remove both cases from further consideration. Note that the original weight was entered in ounces and thus a difference of exactly 100 oz would place both babies in the expected range.

Some of the high values for gestation period are also suspicious, but they are medically possible so we cannot discard them. One of the underlying problems is the imprecision in the measurement of the gestation period. It is often estimated based on the last menstrual period and as such varies with the woman's certainty of the date. Given that the study was performed before the introduction of ultrasound dating, we can assume some variation based on the uncertainty of the gestation period. Some of the factors that are known to influence the birthweight such as gender, parity or geographic location are a priori excluded in this data either by the way the study was conducted or the fact that it is a subset of the full study.

Some categorical variables are directly correlated. A view in the same bin-size mosaic plot readily shows that *Smoking* constitutes a grouped version of the *Quit* variable. Also *Smoking Amount* refines the definition of the *Smoking* variable with respect to the smoking history.

Birthweight vs Gestation

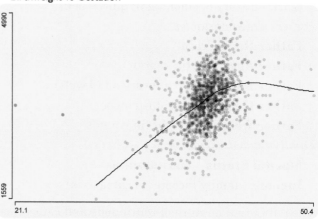

There are two very suspicious outliers with an abnormally short gestation period and yet a normal birthweight.

They are likely data entry mistakes and will be removed from further analysis. The global trend is consistent with the growth of a fetus. Later decline is due to extremely long gestation indicating possible development problems.

Smoking vs. Quit

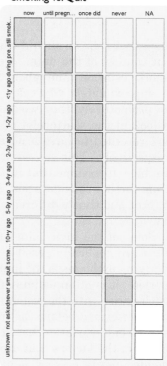

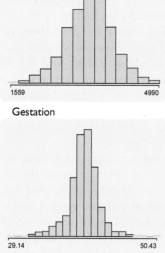

The distribution of both birthweight and gestation are symmetric and unimodal with their respective modi (~ 3400 g and 40 weeks) matching the expected values known from medicine.

Smoking is merely an aggregation of the Quit variable.

Almost all couples were married at the time of pregnancy.

Race (father vs mother)

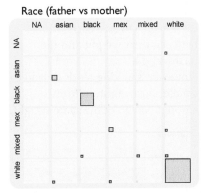

There is a strong trend toward couples of the same race. Exceptions to this rule are often asymmetric.

Age (mother vs father)

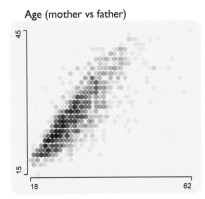

Education (father vs mother)

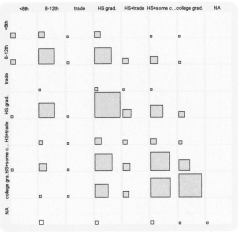

The trend toward a similar level of education is present but not as strong. For couples with more than high school education it is more likely for men to have higher education that their partners which can be further explored using linking in conjunction with the scatterplot of ages.

Analogously the parents of the child are often of similar age. Larger deviations are predominantly seen where younger women have much older partners.

Birthweight by Race

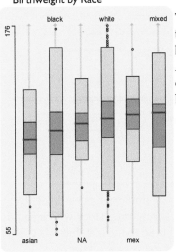

There is a slight trend in the birthweight by race, Asian mothers delivering lighter babies.

The group of women smoking during pregnancy stands out compared to former smokers or those who stopped smoking before pregnancy.

Birthweight by Smoking

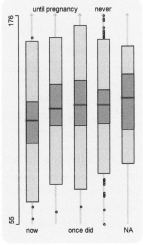

Birthweight vs Mother's Weight

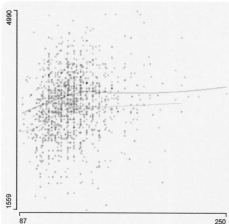

For women lighter than ca. 130 lb. there seems to be a correlation between their weight and the birthweight.

However, smoking tends to lead to smaller birth eight uniformly among women of the same weight.

Smoking

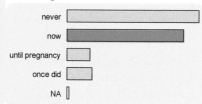

Birthweight

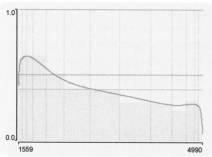

Gestation

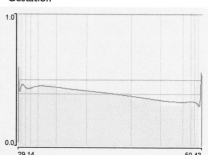

Smokers account for more children with low birthweight than non-smokers.

The gestation time shows a similar but weaker trend.

Education

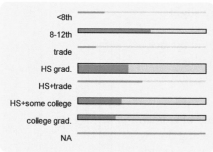

Smoking Amount (in cigarettes per day)

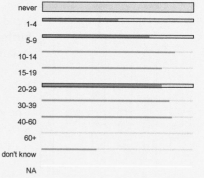

Although the population of some levels is low, there is a slight general trend showing that mothers with higher education are less likely to smoke during pregnancy.

Infrequent smokers are less likely to smoke during pregnancy which could be attributed to the fact that it is easier for them to break the habit.

Both birthweight and gestation period have fairly regular distributions with a clean mode that is in both cases close to the established expected values of 40 weeks gestation and 3.5 kg birthweight. Studies show that the birthweight depends on, among other factors, the gestation period, however, the relationship is non-linear. We can estimate it by using a loess smoother in the *Birthweight* vs *Gestation* scatterplot. The resulting curve is consistent with studies on gestation and birthweight such as Oken et al. (2003). The fetus growth slows down toward the normal birth date and the birthweight starts to decrease for too long gestation periods, indicating complications in the development. Clearly, our data do not have enough cases to support this theory but exhibits a similar trend.

In the next step let us examine some of the descriptive and demographic data. Over 97% of the mothers were married at the time of pregnancy, which may be surprising by today's standards, but the study was conducted among participants of the Kaiser Foundation health plan in the 1960s which likely had an impact on the demographics.

The joint profiles of the child's parents in education and race are best viewed in fluctuation diagrams and the age in a scatterplot. We can see a strong tendency toward same-race couples in fact in about 94% of the cases. The exceptions to this rule are asymmetric: there is no white male with a black female but six black males with white females. Analogously only three white mothers reported an Asian father compared to seven of the inverse combination.

The education of the new parents also shows a pattern of similar education level of both parents. Main asymmetry here is visible among mothers with high school graduation and only some college with college graduated partner. This in part can be explained by looking at the age profile which shows that in many cases the child was born before the mother could finish her degree. Noteworthy in the age profile is also the different age limitation by gender to become a parent, as well as an asymmetric age difference between parents.

So far we have considered interesting patterns in descriptive and demographic variables. The main goal of the study, however, was to analyze the relationship of the birthweight and other variables, in particular the smoking behavior. One way to obtain an overview is to use y by x boxplots with *Birthweight* as the dependent variable. In the boxplot of *Birthweight* by *Race* we can detect a slight trend where Asian women tend to give birth to smaller babies. Interestingly infants of black mothers follow, although the median weight of black mothers is highest among all races.

The boxplot of *Birthweight* by *Smoking* behavior shows a trend for mothers who are still smoking during pregnancy to have smaller babies compared to other groups. However, this does not take into account other factors. Let us review the relationship of mother's weight and the birthweight. By adding loess smoothers we see a pattern of approximately linear relationship that flattens out at ca. 130 pounds. It indicates that up to a threshold the physique of a woman may be a limiting factor for the birthweight of the baby.

We can now select the cases where the mother was smoking during the pregnancy. Comparing the selection to its complement we see a very clear trend that the averaged birthweight is consistently lower for smokers by ca. 250g regardless of mother's weight.

Similarly we can look at the spinogram and a conditional probability plot of *Birthweight* and *Gestation*. Especially the former shows a clear shift where smokers account for most children with low birthweight. The gestation time shows a similar, but weaker trend in the same direction.

Finally, let us analyze the profile of mothers who smoke during pregnancy by looking at spineplots of various descriptive variables. The *Smoking Amount* hints at the fact that infrequent smokers (1–4 cigarettes) are less likely to smoke during pregnancy than heavier smokers. This could be possibly explained by the fact that it may be harder for heavy smokers to quit smoking. The education profile shows a slight decrease of the proportion of smokers with increasing education level, especially if we concentrate on groups with sufficient data.

Further Analysis

In order to determine whether there is a difference in the birthweight between smokers and non-smokers, we can perform Student's *t*-test. First, we should verify that the assumptions are reasonable using a Q-Q normal plot for each group and to check their variance. The birthweight for children of smokers shows a slight asymmetry unlike the group of non-smokers, but both have very similar variance such that we don't have to use Welch approximation to the degrees of freedom. The test can be easily performed in R:

```
> Smokes.Now <- Smoking == 'now'
> t.test(Birth.Weight ~ Smokes.Now, var=T)

        Two Sample t-test

data:  Birth.Weight by Smokes.Now
t = 8.67, df = 1222, p-value < 2.2e-16
alternative hypothesis:
            true difference in means is not equal to 0
95 percent confidence interval:
 196.6378 311.6589
sample estimates:
mean in group FALSE   mean in group TRUE
          3489.098              3234.950
```

The result is highly significant which would lead us to conclude that the birthweight is significantly different between smokers and non-smokers.

Clearly, we are ignoring all other factors involved. In order to reduce the possibility of other factors being the reason for this discrepancy, we can fit an optimal linear model without any smoking indicators and then compare it to a model with *Smoking.Now* added. It turns out that the inclusion of the smoking indicator variable greatly improves the model fit and the variable is highly significant.

Results

In the preliminary exploratory analysis we have detected some data discrepancies in gestation periods that are not plausible. After cleaning those issues we have analyzed bivariate relationships, finding a strong correlation between the race of child's parents and a pattern in their education level.

In the second part we have concentrated on the birthweight which shows relationships with several factors such as gestation period, mother's weight and race but most predominantly smoking behavior. Smoking during pregnancy is a strong indicator for lower birthweight and a weak indicator for shorter gestation period.

Exercises

1. The dataset contains missing data. Is there any pattern in the missingness? How would you analyze this graphically?

2. We have seen a clear relationship between birthweight and mother's weight. Are there any other variables showing a pattern?
   Can those relationships also confirm the theory that smoking leads to lower birthweight?

3. Low birthweight is often defined as a weight of less than 2500g. Which groups are at higher risk of having low birthweight? How would you find them?

4. Does the relationship between mother's weight and birthweight change depending on some other variables? What strategy would you use to find them?

5. Investigate a possible two-way interaction between *Smoking* and *Race* regarding the birthweight, using a boxplot of *Birthweight* by *Race* and the corresponding barcharts of the factors.

   How would you interpret these results?

D

The Titanic Disaster Revisited

Background

The sinking of the RMS *Titanic* was not only one of the worst peacetime maritime disasters in history, but also one of the best known. To date it is constantly popularized by recoveries, books and movies. The historical facts are that the Olympic-class passenger steamship collided during her maiden voyage from Southampton, England to New York City with an iceberg on 14 April 1912 and sank subsequently two hours and forty minutes later at 2:20 am ship time the following day.

The *Titanic* was very advanced, the most luxurious liner of its time and called 'unsinkable' due to its strong design. The extensive media coverage of the disaster can also be traced to the presence of many prominent people of the time on board and famous victims.

The sequence of events on that fateful night is reasonably well documented. The rescue procedure as well as the safety equipment (or lack thereof) have not only led to updated safety rules for ships, but also sparked heated discussion and speculation on why certain people were saved and others were not.

In the following we want to take a closer look at the available data concerning the survival of passengers and the usage of lifeboats. Although the *Titanic* had on board more than the 16 lifeboats required by law, their total capacity was enough to accommodate only about half the passengers present at the time of the disaster.

We will use two data sources for our analysis. The British Wreck Commissioner's Inquiry Report is the source for the breakdown of persons on board by class, gender, age (child or adult) and survival. In addition, we use information on survivors and lifeboats based on survivors data with crew information added. This allows us to follow the events on the fateful night minute by minute.

Study Goals

The official report data are suitable for studying distribution of passengers and crew members on board the *Titanic* and to analyze any patterns that may be relevant to the survival during the disaster. At a global level this should reveal what rules may have been applied when selecting passengers for the lifeboats.

In addition we can follow the sequence of lowering lifeboats and the relationship of their location and time with respect to the people on board. This more precise information allows us to study spatial and temporal patterns.

Description of Data

The survival dataset is a direct reproduction of the numbers mentioned in the British report. It consists of 2201 cases and four variables:

Class - class (first, second, third, crew)

Age - dichotomized (child, adult)

Gender - gender (male, female)

Survived - survived (yes, no)

In the collapsed form the dataset also contains the *Counts* variable which is then used as weight for each category combination (cf. Section 2.4).

The lifeboat dataset is a subset of the above and contains counts of surviving passengers only as well as variables describing the lifeboat that the passenger boarded. The variables are as follows:

Class - class (first, second, third, crew)

Gender - gender (male, female)

Boat - number of the boat (1–16, A–D)

Sequence - number of the boat in the launch sequence

Side - location of the boat (starboard, port)

The *Sequence* and *Side* information is derived from the boat number. The dataset is compiled from data available on the surviving passengers collected from various sources.

Graphical Analysis

First let us look at the official survival data. The basic information on the passengers of the *Titanic* can be viewed easily with barcharts. We can see that the vast majority of the passengers were adults and there were by far more men than women. The crew formed the largest group, closely followed by the third class. Perhaps surprisingly the number of second class passengers was only slightly fewer than first class.

We can see that only about a third of the passengers survived. By selecting the survivors, we can track the absolute number of survivors in each category. We can see that about the same number of men and women were saved despite the gender imbalance overall. Switching to spine plots will show us the survival rates immediately. We can see the survival rate decline by class and a huge difference between men and women. When we remember that the *Titanic* did not have enough lifeboats for all passengers, it supports the 'women' part of the general rule "women and children first."

As a secondary rule it appears that the class of a passenger was taken into account (directly or indirectly by the location of the quarters for each class). To verify this behavior let us look at the relationship of gender and class in a mosaic plot, using highlighting for survival. It becomes apparent that a vast majority of women in the first class, second class and crew were saved, not quite so in the third class. The men, however, had far worse chances of survival, with the crew surpassing even the second and third class in survival, mostly due to the fact that each boat had to have a crew member on board for the operation of the lifeboat.

Previously we have seen a preference for women, but the trend was weaker for the children as seen in the spine plot. In order to have a closer look, let us look at a same-bin mosaic plot for *Class*, *Age* and *Gender*. This reveals that all children in the first and second class were saved whereas the third class was left behind, suggesting that the rule may have been "women and children in the first and second class first." This is, however, just a speculation, but we can use our second dataset on

Survived

Only about a third of the passengers survived.

Class

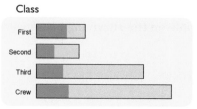

Crew was the largest group on board, followed by the third class.

Class

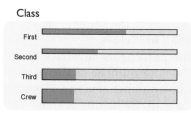

Survival rate declines with the class.

Gender x Class

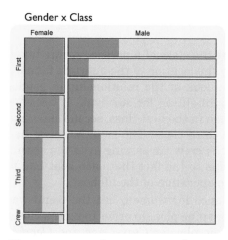

There were very few women in the crew. Almost all women survived except for the third class. Men in the first class had the best survival chances.

Age

The proportion of children on board was quite low.

Gender

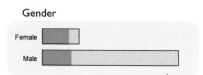

Although there were far more men on board, the gender was about equally distributed among the survivors.

Age

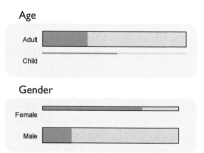

Gender

"Women and children first" seems to hold true at least for the women.

Class x Age x Gender

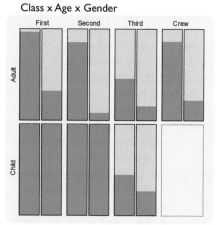

Within each class the preference toward women and children is clear.

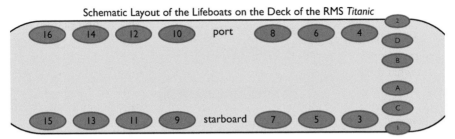

Schematic Layout of the Lifeboats on the Deck of the RMS *Titanic*

Boats 3 and up were full-size with a capacity of 65 people each, A–D were slightly smaller, collapsible versions with a capacity of 47 and finally, boats 1 and 2 the smallest with the capacity for 40 people.

More passengers were saved on the port side which also saved more men.

The inequality in gender proportions stems from the fact that men were apparently prevented from boarding on the port side.

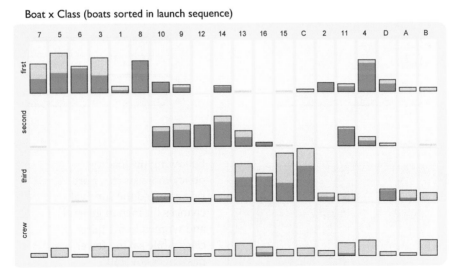

Boat x Class (boats sorted in launch sequence)

The launch sequence shows clearly a loading by class. Crew members were needed to operate of the lifeboats and were therefore present on all boats.

Most lifeboats were
loaded with many
fewer passengers
than what their
capacity allowed.

The loading pattern by
class over time is clearly
visible. It is consistent
until 1:45 am (boats 2 and
11) when the loading
becomes more erratic.

Class

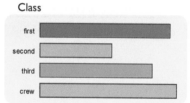

Boat

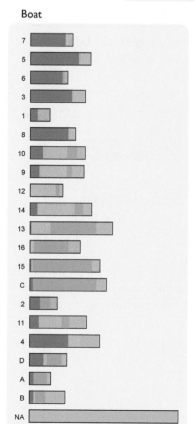

Boat

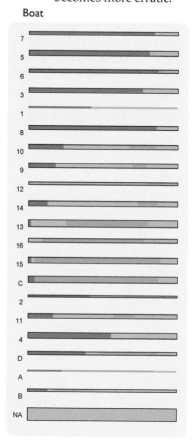

Gender x Side

Following the loading
policy more strictly, port
side reduced the survival
chances for men in general
and women in the third
class while saving fewer
passengers overall.

surviving passengers to learn more about the sequence of events as it also contains the temporal information that we are missing here.

The boat launch data do not contain the age, but otherwise the same classification as the official data. In addition, however, it features the boat number for each passenger. First let us review the boat layout on board of the *Titanic*. They were located on each side of the vessel (port side is the right-hand side when looking toward the ship from the front; starboard side is the opposite side). On each side there were seven full-size lifeboats, one smaller boat and two collapsible boats. The boats on the port had even numbers, the collapsibles were named by letters A–D.

First, let us compare the profile of each side as they were under the command of different officers. Given that the ship did not turn on its side while sinking, we would not expect large differences. The *Side* barchart shows us that more passengers were saved on the starboard side. To account for the passenger class we can use multiple-barcharts mosaic plot of *Side* and *Class*. It is somewhat striking that many more third class passengers were saved on the starboard side than on the port side. We can also see that all missing values are due to the crew — we have boat information only for a fraction of the crew members.

We can add gender to both plots using highlighting, selecting the female passengers. The loading officers on the port side have apparently stuck much more closely to the rule "women and children first" as well as loading by class. Note that we do not have the age information and therefore it is likely that the men shown here may be children. According to some survivors, some men tried to impose as women by wearing a wig to increase their chances to get on a lifeboat, supporting our findings.

Our data also includes the time of the launch for each boat, allowing us to look at the sequence in which the boats were lowered. The easiest way to sort boats in their launch sequence is to plot a same bin-size diagram of *Sequence* by *Boat* along with a barchart of *Boat* and draw boat numbers in the barchart according to the columns in the diagram to obtain a diagonal line.

With the boats sorted in the launch sequence, we can look at the temporal aspect of the data, for example, by using a multiple-barcharts plot of *Boat* by *Class*. We can clearly see that lifeboats that were launched first (first boat was launched at 0:45, more than an hour after the impact) contained only first class passengers along with some crew members. Later (boat 10 which was launched at 1:20) second class passengers started to appear. We can also see that crew members were present in all boats, confirming that they were intended for the operation of the boats.

Let us further look at the utilization of the boats by plotting a barchart of boat numbers. It is evident that boats were almost always loaded with fewer passengers than what their nominal capacity allowed. Boat number 13 is the one with most surviving passengers, sporting only 51 passengers. That is still less than the official capacity of 65 souls and yet it was the fullest boat. Boat number 1 became especially notorious as its only passengers were seven crew members and five first class passengers, three of which constituted the party of Sir Cosmo Duff-Gordon, who was accused of bribing the crew.

We can add class information by using the *Class* barchart to brush each class with a different color. Since the boats are sorted in launch sequence, we can see a pattern in the proportions within each boat. As we are interested in the proportions it is better to switch to a spineplot of the boats. The pattern becomes very clear, showing the shift in classes over time up to boat 2. From there on, it seems that the loading become more hectic as the pattern first-second-third class repeats again, possibly serving passengers that did not make it in time for the first round.

Finally to have a look at the differences between loading patterns at the sides of the *Titanic*, we can use a *Gender* and *Size* mosaic plot along with the color brushing. It confirms that women had higher priority at the port side and that class membership was observed more strictly.

Results

The analysis of the official report as data source gives basic facts on the distribution of passengers on board the *Titanic*, such as the lack of children, the few women among the crew members or the fact that the second class was the smallest. It also shows that the survival rate decreases with the class and that the "women and children first" rule seems not to apply as rigorously to the third class passengers — almost all women and children of the first and second class were saved which cannot be said about the third class.

The analysis of the boats and the launch sequence suggests that lifeboats were at first filled according to the class and often undermanned, which is quite surprising given that the *Titanic* had enough boats only for about half of the people on board. The loading procedure was also different on each side of the ship with the port side overcommitted to the rules yet saving far fewer people.

Analyses are only as good as the numbers they are based upon and the exact numbers of the *Titanic* survivors by class are disputed up to this date. Although some trends were very clear in the analysis, it is not uncommon to find very different "*Titanic*" datasets. Especially the assignment to lifeboats is ambiguous because some boats picked up passengers floating freely in the water or from other boats. Quite often some details are imputed or assumed which can be very far from the truth. Dataset versions have been seen where all missing values are attributed to women, creating very different (and false) results.

The lessons learned from the disaster helped improve the safety regulations for generations even though what happened exactly that fateful night remains a mystery, open for speculation and tall tales. Even the discovery of the shipwreck failed to shed more light on the incident but instead recalled century-old memories.

Exercises

1. Determine the survival rate for all subgroups using the official data. Which plot would also allow you to see the absolute numbers at a glance?

2. Assuming that the survival rate depends only on the gender, are there significant differences in the survival rate of the classes?

3. White Star Lines claim that there was no difference in the loading procedure on each side of the *Titanic*. Can you verify this claim statistically?

E

Housing Rent Prices in Munich

Background

It is rather common to rent an apartment or a house in large cities in Europe. In Munich, Germany, for example, the majority of households live in rented quarters. Munich is the third largest city in Germany, located in the southern part of Bavaria, close to the Alps. It is only natural that an important question for a tenant is whether the rent he is paying is comparable to similar properties in the area.

In Munich the question is also important for the entire business of housing tenure, because laws govern the amount that a lessor can demand as a rent: the lessor can only increase the rent if *"the current rent does not exceed the commonly accepted rent for a housing comparable as of size, equipment and location."* It is thus understandable that lessors are interested in quantifying *"commonly accepted rent"* for their type of housing as high as possible, wheres the lessees have the opposite interest.

In order to resolve this conflict, the city of Munich performs a survey every two years to objectively determine the average rent for a given area and given kind of dwelling. Due to the size of the city, it is not feasible to collect data from every single household so the survey is performed via a telephone poll. The resulting data are then used to fit a model which allows the calculation of the *"commonly accepted rent"* for any housing property in Munich.

Given that billions of Euros are involved in the housing tenure business each year in Munich alone, it is not surprising that the housing rent survey sparks heated discussion each year.

Study Goals

The primary goal of the study which is the source of the data was to estimate the average housing rent prices in Munich taking into account the size, location and amenities of the dwelling. Costs of utilities (such as electricity, gas etc.) were not included in this study. However, the data are rich providing information on the properties themselves as well as the geographical location (district) and rent information. This allows us to study the relationship between the age of the housing, its size and location in addition to the original question of rent development.

In the first part we want to assess the quality of the data, then consider relations and geographical aspects and finally look at the most influencing factors for the rent.

Description of Data

The dataset consists of 2,053 observations and 14 attributes where each observation corresponds to one survey respondent and thus one dwelling. The publicly available dataset is a subset of the data collected for year 2003 which consisted of 3,118 qualified entries.

Dataset Variables:

Rent - rent in Euros per month

Size - size of the dwelling in m^2

Num. Rooms - number of rooms (excludes kitchen and bath)

Built - year in which the construction of the building was completed

District, District No. - name and number of the district of Munich in which the dwelling is located

Neighborhood - *average, good* or *best* as defined by the city of Munich based on the location

The following variables are all binary (yes/no):

Warm Water - availability of in-house warm water

Central Heating - availability of central heating

Tiled Bath - tiles in the bathroom

Plus Bath - additional, unusual amenities present in the bath

Plus Kitchen - kitchen featuring additional amenities

Graphical Analysis

The first step is to review all variables using univariate plots. The data does not contain any obvious errors, although the histogram of the *Built* variable hints at some discretization in older properties. Apparently the

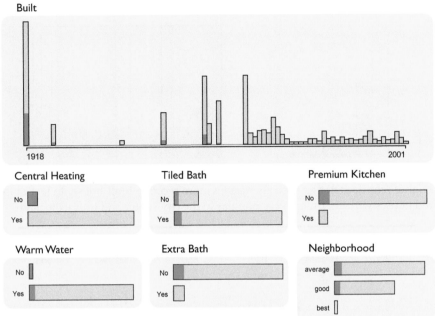

Histogram of the build date reveals censoring and possible jittering for some value ranges.

Barcharts of all categorical attributes in the data give us an idea of their distribution. Selected are dwellings that have no central heating or no warm water. Clearly those are all old buildings.

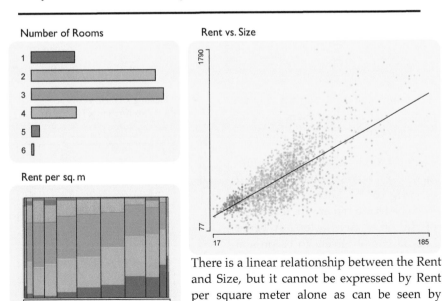

There is a linear relationship between the Rent and Size, but it cannot be expressed by Rent per square meter alone as can be seen by brushing the number of rooms.

Rent

Built

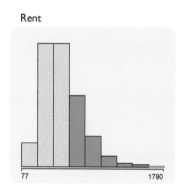
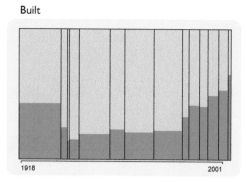

77   1790

1918   2001

Higher rent does not necessarily mean a more recently built house. In fact quite old houses with charm are very popular in the city center and thus expensive.

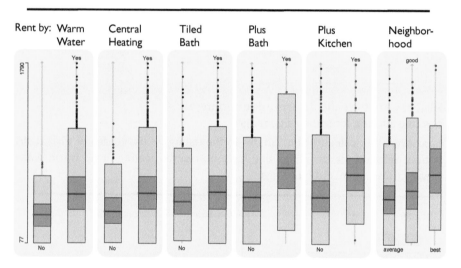

Rent by: Warm Water — Central Heating — Tiled Bath — Plus Bath — Plus Kitchen — Neighborhood

Amenities justify a slightly higher rent in general, but there are still many exceptions.

Central heating is almost a require-ment for warm water; the vast majority of the houses have both. Extra baths are rather rare regardless of the other two factors. A better kitchen is more likely to be present with a better bath.

The majority of dwellings, however, have all basic amenities, but none of the luxurious ones.

Central Heating x Warm Water x
Plus Bath x Plus Kitchen x Tiled Bath

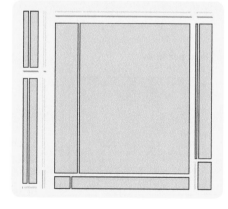

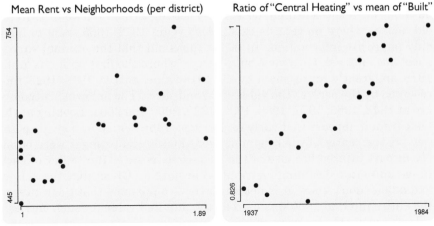

Mean Rent vs Neighborhoods (per district)

Ratio of "Central Heating" vs mean of "Built"

Aggregated statistics per district show a relationship between mean rent and the average quality of the neighborhood. The lack of central heating can be traced to the age of the buildings.

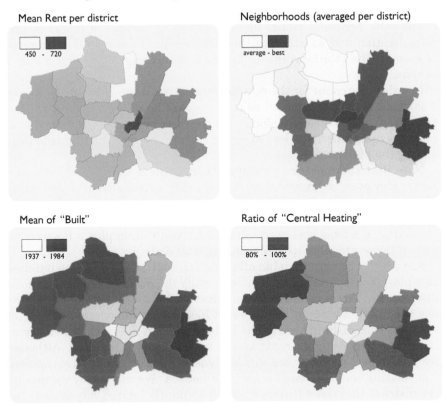

Mean Rent per district

450 - 720

Neighborhoods (averaged per district)

average - best

Mean of "Built"

1937 - 1984

Ratio of "Central Heating"

80% - 100%

There are clear spatial trends in the housing properties and the rent. The old city center is apparently very popular despite old buildings and limited amenities.

data has been left-censored, possibly by setting the build date to 1918 for old houses where no records exist. Also years 1918–1965 seem to have only imprecise information. In fact it turns out that the original survey sheet does not ask for a year but a range, where the first option is 'until 1918' apparently represented by the 1918 value, next is '1919–1929' represented by the peak at the value 1924 and so on. The following cutpoints are at 1948, 1966, 1977, 1988, 1997, 1999, 2000 and 2001. Looking at the histogram it appears that early ranges were represented by a single point (mostly the center of the range) whereas more recent ranges were possibly in part jittered because of the suspicious peak at the center of each range and often declining distribution around it. Given that the survey had a field 'don't know' it is not far-fetched to presume that the center of the entire interval 1960 was used as the imputed value for such entries.

In the next step we want to have a closer look at the response variable *Rent*. We show its relationship to the size of the housing in a scatterplot. Clearly, we see a strong positive correlation of the rent and the size. The relationship is mostly linear and the corresponding least-squares linear regression line is shown in the plot.

The color brushing allows us to include the number of rooms in this consideration and as expected the increase in size is mainly to number of rooms. The relationship can be more closely explored using a x-by-y boxplot.

In order to account for the relationship between the size of the housing, it is usual to use rent per m^2 as a measure of how fair a given rent is. A spinogram of this measure along with the color brushing of the number of rooms reveals a perhaps less expected result: one-room apartments have among the highest rent per m^2.

Note that the measure corresponds to a linear model without intercept. However, a linear model based on the data (easily fitted using lm(Rent˜Size) in R, or by querying the regression line in Mondrian) shows an intercept of 89.85 Euro. This can be interpreted as the rent consisting of a fixed cost associated with the dwelling plus a linear component proportional to its size (roughly 6.90 Euro per m^2).

Size is not the only variable affecting the rent. The age of the building also plays an important role. Unfortunately the *Built* variable is partially discrete, making a comparison in a scatterplot hard. We can, however, use the linked view of a histogram of *Rent* and a spinogram of *Built* (or vice versa). Brushing bars in the histogram from the right tail shows us the proportion of higher rents among various ages of building. Although we may expect rising prices with increasing date of construction, we see a surprising rent increase for very old houses. We can see by comparing the districts that those houses are predominantly located very close to the city center and as such very popular.

Further variables that may increase the popularity of housing are amenities which are coded as binary variables in the dataset. We can review their effect by looking at boxplots of *Rent* conditional on each categorical value. In general we see a trend: the presence of an amenity leads to a slight increase in the rent. The same holds true for better neighborhoods. However, it is not clear whether the influence is due to each factor independently or merely the manifestation of the same effect.

We will need to look at the joint distribution of the categorical variables to answer this question. The easiest way to explore possible interactions is to use mosaic plots. Building them in a bottom-up fashion interactively allows us to look for patterns. For example, it is evident from a mosaic plot of *Central Heating* and *Warm Water* that those are dependent — the availability of central heating is a strong indication for the availability of warm water, which can be explained by the fact that most central heating systems in German cities use water circulation. More than 80% of all households have both warm water and central heating.

Adding further interactions shows us that special features are more likely to occur together, although they are not common in many apartments. The only exception is tiled bathroom which is quite common - about 63% of all houses have warm water, but no extra amenities save for a tiled bathroom.

Finally, let us consider the spatial aspect of the data. Among the 25 districts of Munich the top 3 with respect to the median rent are *Trudering – Riem*, *Altstadt – Lehel* (old city center) and *Bogenhausen* which can be seen from the a boxplot of *Rent* by *District*. In order to place this information in context, we can aggregate all variables by district and display corresponding choropleth maps. The *Trudering – Riem* district in the eastern part of Munich features the most recent buildings on average and is also considered a very good neighborhood. *Altstadt – Lehel* is in contrast one of the older parts, but its proximity to the city center makes it very attractive, despite the lack of some amenities in older buildings.

Notable also is that *Maxvorstadt*, which is on average the best neighborhood, does not excel in any of the categories with the exception of lack of central heating.

Further Analysis

The estimation of the rent with a linear regression model was proposed in the literature for the original dataset. Our graphical analysis showed that we may achieve reasonable results using linear models, because *Rent* showed quite a strong linear relationship with *Size*. We can start with a very simple linear model with only one independent variable *Size* as mentioned in the previous section.

By selecting high and low residuals (setting anchor point to 0 and bin width to σ) we can see in Fig. E.1 that the rent for newer houses is underestimated and for older houses overestimated. Thus we can add the

Residuals of a LM Rent = Size + ϵ   Built

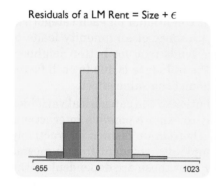 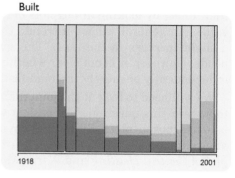

FIGURE E.1

Histogram of the residuals of the linear model *Rent=Size+ϵ* and spinogram of *Built*. Residuals $> \sigma$ are brushed in red and $< \sigma$ in blue.

interaction of *Built* and *Size* to the model and obtain a better fit. The model can be expressed in R as follows:

```
> m<-lm(Rent ~. + Size*Built -District -Rent.per.sqm, rent)
> summary(m)
...
Coefficients:
                      Estimate  Std.Error  t value  Pr(>|t|)
(Intercept)           1.45e+03  8.21e+02    1.774    0.0762
Size                 -4.25e+01  1.00e+01   -4.247  2.26e-05
Built                -8.45e-01  4.20e-01   -2.012    0.0443
Num..Rooms           -3.32e+01  6.42e+00   -5.172  2.54e-07
Neighborhoodgood      5.38e+01  6.99e+00    7.706  2.01e-14
Neighborhoodbest      1.44e+02  2.29e+01    6.269  4.43e-10
Warm.WaterYes         1.63e+02  2.12e+01    7.684  2.37e-14
Central.HeatingYes    7.39e+01  1.45e+01    5.097  3.76e-07
Tiled.BathYes         4.29e+01  8.65e+00    4.957  7.73e-07
Extra.BathYes         4.87e+01  1.20e+01    4.067  4.94e-05
Premium.KitchenYes    1.11e+02  1.30e+01    8.514  < 2e-16
Size:Built            2.58e-02  5.15e-03    5.018  5.68e-07
...
```

```
Residual standard error: 150.1 on 2041 degrees of freedom
Multiple R-Squared: 0.6281,      Adjusted R-squared: 0.6261
F-statistic: 313.4 on 11 and 2041 DF,  p-value: < 2.2e-16
```

Adding *District* as a factor yields no significant improvement over the above model. Interestingly, the original report was modeling the rent per m^2, which is problematic, as we have seen, due to the relationship of rent and size.

Results

We have discovered some undocumented manipulation in the *Built* variable that affects recent values. The housing amenities depend on the age of the building as expected, but there is a "standard" set of amenities available in the majority of houses. The rent amount per month is predominantly directly proportional to the size of the housing. Other factors can positively influence the rent such as proximity to the city center or location in one of the good neighborhoods, although not exclusively.

The spatial profile of the expansion of Munich is clearly visible even when aggregated at the district level. The same applies to amenities such as central heating, but not necessarily the rent.

The Munich rent dataset is rich in continuous, categorical and spatial data. We learned not only about the relationship of the rent prices to all other variables, but also about the history and different neighborhoods of Munich.

Exercises

1. The mosaic plot of *Central Heating*, *Warm Water*, *Plus Bath*, *Plus Kitchen* and *Tiled Bath* on the graphics page showed the interaction of those amenities. How would you compare the three neighborhood types in this respect? Are there any differences?

2. Not all combinations of the binary variables occur in the dataset. How would you look for such missing interactions? Find them for combinations of two and three variables respectively.

3. Compare the districts with respect to various variables. Are there surprising outliers? What plot types would you use?

4. Fit a linear model using both *Size* and *Built* variables and include the residuals in the dataset. How do they differ from the residuals of the simple *Size* model with respect to other variables?

F

What Makes a Tour de France Winner

Background

The Tour de France is the world's longest, most watched and probably the most well-known cycling race. According to the Times, it is "arguably the most physiologically demanding of athletic events." It consists of approximately 20 stages in France with occasional visits to neighboring countries. The Tour started more than a century ago in 1903 and it became very popular, increasing the popularity of cycling as a sport.

FIGURE F.1
Armstrong and Basso 500m from the mountain arrival in La Mongie at the 2004 Tour de France.

A typical Tour de France is about 3500km long and is divided into "stages", which are races held over one day. Regular stages are started with a mass start. However, to avoid dangerous mass sprints, all riders in an identifiable group at the finish line share the finish time of the leader of such a group. Special stages include an *individual time trial* where riders are started sequentially and a *team time trial* where several riders of the team share the finishing time.

In this chapter we take a closer look at the 92nd Tour de France of 2005 which consisted of 21 stages with a total length of 3,607 km. The route started in the small town Fromentine on the Atlantic Coast on July 2 with an individual trial and finished traditionally in Paris on July 24. A total of 189 riders in 21 teams of 9 competed. Lance Armstrong won the overall classification for the seventh consecutive time, concluding his exceptional carrier as road racing cyclist.

Study Goals

The Tour de France data allow us to illustrate the analysis of specific longitudinal data, consisting of stage times and ranks, in conjunction with descriptive data on riders. One goal is to highlight patterns in the times that are due to different properties of the stages as well as interactions of riders by their role and team. In addition, temporal dependence between stages with respect to riders is of interest. Given the relationship of individual times to overall classification times and ranks, it allows us to consider the influence of individual short-term events on the whole race. In addition we will try to understand different roles of riders and teams from their performance.

Description of Data

The Tour de France 2005 data are organized in a matrix such that each row corresponds to a rider. The columns consist of descriptive attributes such as name, team, nationality, year of birth, role and longitudinal data across stages such as stage time, cumulative time and rank in the general classification after each stage as follows:

Name - name of the rider

Team - name of the team

Nationality - nationality of the rider

Year of Birth - year of birth of the rider

Type of Rider - classification of the rider

Sn - stage time (for nth stage) in seconds

Tn - total (accumulated) time up to nth stage (inclusive) in seconds

Rn - rider's rank in the general classification after the nth stage

Variables containing the stage number also include a suffix to describe special stages such as individual time trials (Time), team time trials (Team) and mountain stages (Mount.). Riders who dropped out are identified as missing values in stage times and ranks for stages in which they no longer participated.

Graphical Analysis

Descriptive attributes of the dataset are easily viewed with barcharts and histograms. The riders were of 21 to 37 years old with an almost uniform distribution except at the extremes. Larger European countries where cycling is historically popular such as Spain, France, Italy, Germany and Belgium provided most riders, followed by Australia and U.S. where the popularity outweighs the distance. The size of a team is constant and

Year of Birth

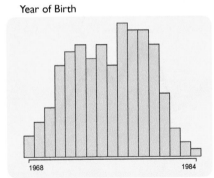

1968   1984

Riders' ages range between 21 and 37, with an almost flat distribution between 24 and 33 years.

Nationality

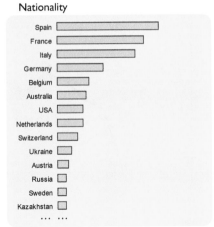

Spain
France
Italy
Germany
Belgium
Australia
USA
Netherlands
Switzerland
Ukraine
Austria
Russia
Sweden
Kazakhstan
··· ···

Spain, France and Italy contribute almost half of all riders.

Year of Birth by Team

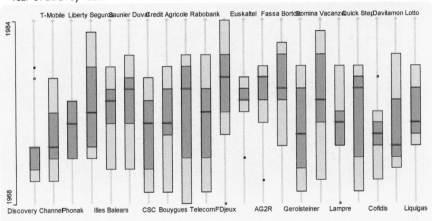

Total time by Team

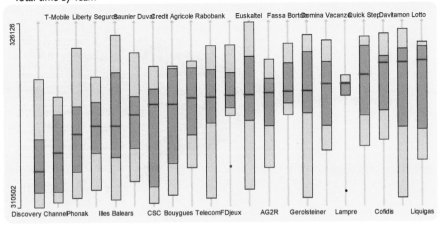

Stage Time

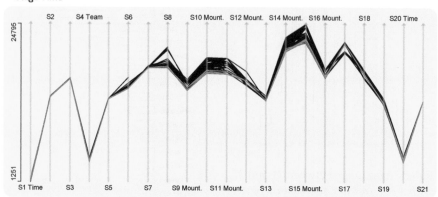

Ranks

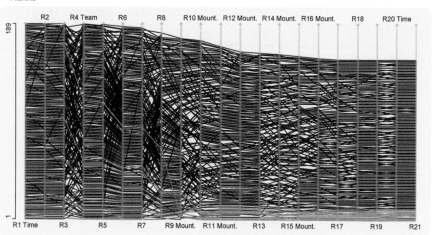

Accumulated Time

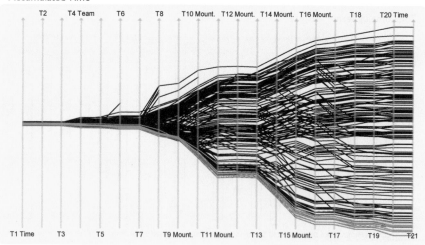

Stage Time

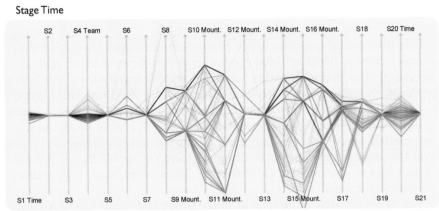

The team of the winner (in red above) managed to have at least one helper with the leader most of the time, which cannot be said for most other teams.

Final Rank by Rider Type

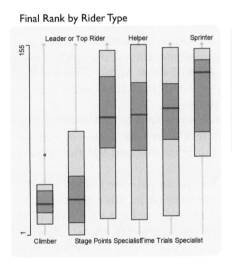

Rider Type

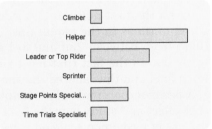

Climbers do very well in the general classification as opposed to other specialists.

What makes a winner (highlighted below)? Be among the top riders in the mountains where it matters and stay in touch with the main field.

Stage Time

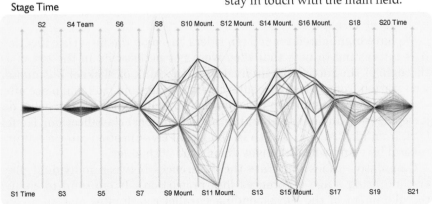

each row corresponds to a rider, thus the corresponding barcharts simply confirm that the data are as expected.

One interesting question is the composition of teams as of riders' age. Boxplots *Year of Birth* by *Team* give us some insight as to which teams can be classified as 'young' or rather 'experienced.' We can see that even teams with many young riders such as *FDjeux* have at least one older, experienced rider. Interestingly enough, experience may play a role, because leading teams such as *Discovery Channel* and *T-Mobile* have the highest median age. However, age itself is not the only indicator as *Cofidis* and *Davitamon Lotto* show. Both teams have rather old riders but rank almost last in the boxplot *Total time* by *Team* when sorted according to median *Total time*.

Longitudinal attributes such as stage times and ranks are best viewed in parallel coordinate plots. A global profile of all stages on a common scale highlights the duration of stages with respect to each other, but also allows a comparison of the time differences within a stage relative to stage times. It is evident that in the first five stages the time difference between the first and the last rider is essentially negligible when compared to times in the mountains. A complementary scatterplot of stage times (e.g., between 2nd and 3rd stage) will show highly discrete stage times due to the fact that most riders (183 in that case) arrived in the main field and thus shared the same time.

A parallel coordinate plot of ranks allows us to remove the time difference effect and study how the field changes after each stage. In addition, the use of a common scale on ranks highlights dropped-out riders as the highest ranks become vacant.

We can see mostly horizontal lines for stages with very small effect on the global classification. First major changes occur after the team trial which is due to the effect of a good team time pulling all riders of the team forward even if they are usually not in the top spots. However, we know from the stage time plot that this effect was rather temporary due to the small absolute time difference. Further major changes in the classification are visible in the mountains. The overall ranks change very little in the last stages.

We have highlighted the top ten riders based on the final classification. It is interesting to note that they were not necessarily in the top place throughout the race. Most interesting is the rank of Mickael Rasmussen who started from 174th place after the first stage and took 2nd place after the 10th stage. Rasmussen is a climbing specialist and won the polka dot jersey as the King of the Mountains (best climber) classification. He fell back from his 3rd place to 7th in the global classification after a fall and several bike and wheel changes during the 20th stage.

Cumulative times illustrate the differences between individual players and bring stage differences into proportion. It is useful to change the alignment of the corresponding parallel coordinates plot on a common scale to highlight relative differences. Difficult mountain stages very clearly spread the global classification apart. It also brings the results of early stages into the right perspective, explaining how spectacular shifts in ranks such as Rasmussen's are possible. Among the top ten riders only Óscar Pereiro Sio stands out in this plot and was able to improve his rank substantially despite a suboptimal performance in the first mountain stages. He achieved this by initiating breakaways in stages 15 (toughest on the Tour), 16 and 18 and was awarded the Most Aggressive Rider Award.

The same plot for individual stage times highlights the contribution of each stage to the global classification. It is also useful to assess the stage-wise performance of individuals or teams. Selecting different teams shows fairly high variability in each team. Teamwork is important in road cycling as teammates can help the team leader by shielding him so that he can ride in the slipstream or by preventing breakaways of other teams. The *Discovery Channel* team did a good job helping the winner Lance Armstrong by having teammates in his proximity. Other teams were not as successful.

Different rider types have different roles or tasks. As expected, we can see in a boxplot of the final rank by rider type that leaders finish before most other rider types. Interestingly, climbers show very similar performance. Due to the time gains that can be achieved by breakaways in the mountains, they tend to perform well in the global classification. Rather surprisingly sprinters form the least successful group in the global classification.

Further Analysis

One interesting question is whether individual stage performances can be indicators to predict the global classification outcome of the race. The assessment is complicated by the rule that the main field shares the arrival time and thus some stages show no variance for most riders. Interestingly the first time trial shows basically no correlation (cor=0.29) to the final outcome. In contrast the final time trial is more highly correlated (cor=0.70) with the final result. However, for all practical purposes that is too late as the cumulative general classification is unlikely to change given the influence of the last stage.*

Hence a related question is: after how many stages is the race result unlikely to change much? Again, using the simple tool of correlation we can compute the correlation between the final result and the cumulative time after each stage. This can be done with the following R code:

---

*It is actually an unwritten rule that there are no more attacks during the last stage that might change the classification.

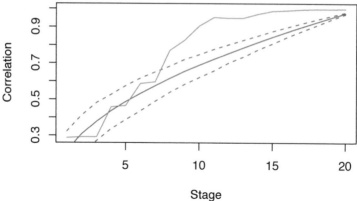

FIGURE F.2

Correlation of cumulative stage results to the final time (data in red, baseline in gray).

```
> st <- TDF[,27:47]              # cummulative times
> st <- st[!is.na(st[[21]]),]    # filter out drop outs
> # calculate correlations for stages 1..20 and plot them
> sc <- unlist(lapply(st[,1:20], cor, st[[21]]))
> plot(sc, col=2, type='l',xlab='Stage',ylab='Correlation')

> sim <- function(...) {         # simulation function
+      m <- matrix(runif(prod(dim(st))),21)# random times
+      cm <- apply(m, 2, cumsum)           # cumulat. times
+      apply(cm, 1, cor, cm[21,])[-21]     # correlations
+ }                                        # excl. stage 21

> s=matrix(unlist(lapply(1:500, sim)),20) # simulate 500x
> lines(apply(s, 1, median))
> lines(apply(s, 1, quantile,0.025), lty=2)
> lines(apply(s, 1, quantile,0.975), lty=2)
```

Fig. F.2 shows the results. As a baseline for comparison, the gray line is the correlation for (uniformly) random stage results along with a 95% confidence interval. It is evident that the first mountain stages contribute heavily to the final classification.

Results

What makes a Tour de France winner? A stage profile shows that Lance Armstrong showed the overall most solid performance without any apparent weaknesses. He could consistently keep up with the top riders, especially in the first part of the tour, but did not waste energy on trying to win every stage or follow every breakaway. Yet he always stayed in touch with the main competitors.

Exercises

1. We have seen some trends in the composition of teams with respect to the age of the riders. Are any effects visible at the individual level?

2. How would you analyze which stages generated the most dropouts during the race?

3. Which type of rider is most likely not to arrive at the Champs-Élysées in Paris, but to drop out at an earlier stage?

4. The baseline in Fig. F.2 was computed by assuming random stage times. This disregards the actual distribution of times. Calculate the baseline by randomly permuting the stage times for riders in a stage. How are the results different from the previous approach? Is the result expected? How can you interpret it?

How to Survive the Thirty Years' War

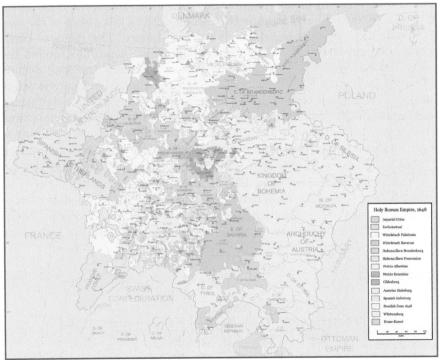

FIGURE G.1
The Holy Roman Empire in 1648, scattered into dozens of small realms.

Background

The Thirty Years' War was fought between 1618 and 1648 on mainly German territory. It was a succession of many "smaller" wars and the first war that involved most of the major European continental powers. On the other hand, it was the last major religious war in mainland Europe.

Although religious conflicts between Protestants and rising Calvinists on the one side and Catholics on the other side were the basis of the war, the political aspect cannot be neglected. Spain, which held some Dutch territories, and the Holy Roman Empire were both part of the Habsburgian realm, surrounding the French Kingdom, which in turn had an interest in gaining power over the weaker German states. Denmark and Sweden shared an interest in expanding their power over the surrounding states of the Baltic Sea, mainly the northern German territories.

Religious tensions were high in the early 17th century. It only needed the spark of the Prague Defenestration of two Catholic councillors in May 1618 to ignite the war by the Bohemian Revolt. It soon spread over all of central Europe, starting the Thirty Years' War. The Bohemian Revolt led over to the Palatinate phase from 1621 to 1625. Then followed periods of interventions of the neighboring states, i.e., Denmark 1625–1629, Sweden 1630–1635 and France 1636–1648. In the last years of the war the combatants sought negotiations more and more. The Peace of Westphalia, signed October 24, 1648 in Münster and Osnabrück, ended the war. Besides the reprisals of the Westphalian treaty, the long decline of the Holy Roman Empire began as Germany then was a patchwork of mostly independent territories (cf. Figure G.1). Whereas the neighboring countries soon formed sovereign nation-states with rising economies and well controlled national armies, it took another 200 years for Germany to gain imperial power, resulting in the catastrophes of the 20th century.

Several of the generals of the involved troops (almost half a million on both sides, including more than 100,000 Germans each) are still famous today. Tilly and Wallenstein for the Catholic League and Gustav Adolph II and Cardinal Richelieu for the Protestant Union fought famous battles and lost their lives on the battle grounds (except for Richelieu).

FIGURE G.2
The people of Augsburg swearing support to King Gustav Adolph II.

Civilian suffering and losses reached unprecedented numbers during the war. An estimated 3-4 million casualties amount to roughly 20% of the 17 million population of the Holy Roman Empire at that time. Most of the destruction was caused by the unparalleled cruelty and greed of mercenary soldiers, and as a consequence, both typhus and dysentery had become endemic in Germany in the last decades of the war.

Goals of Study

Whereas the historic background of the Thirty Years' War had been well investigated, the influence on the socio-economic structures of that time have been far less explored. Looking at census and tax data of the city of Augsburg before and at the end of the war will give insights on how the socio-economic structure changed for this city during the Thirty Years' War.

As Figure G.2 shows, historians know very much about kings (such as Gustav Adolph II), but only very little about the people (of Augsburg).

Description of Data

The dataset is based on tax data from historical tax books of the city of Augsburg in the years 1618 and 1646.* The data on 8,748 taxpayers in 1618 and 4,691 taxpayers in 1646 was then aggregated to 95 tax districts based on a map of tax districts dated 1626. For each of the districts the following data were recorded:

Name of the District

Number of Taxpayers in 1618 and 1646

Tax Payed in 1618 and 1646 in Gulden

Percentage of Major Professions - % of merchants, weavers and socially subsidized workers. If a profession was not found at all in a district, it was recorded as NA.

Percentage of Catholics in the district

Average Age

All variables without a specific date state the situation in 1618. To give a better understanding of the geographical aspects of the data, a map of the 95 tax districts is linked to the data.

Graphical Analysis

The graphical analysis will be threefold. We first look at the socio-economic structure of the city of Augsburg before the war. Then we investigate the geographical aspects of the data. Finally, we look into the tax information at the end of the war and try to find out how much of a structural change took place.

In the scatterplot of *Tax 1618* vs *No. Taxpayers 1618* we find a typical L-shape, showing that only very few tax districts contribute to most of the

---

*Although the war did not end "officially" until 1648, there was not much of a change during the last two years.

Tax 1618 vs. No. Taxpayers 1618

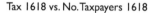

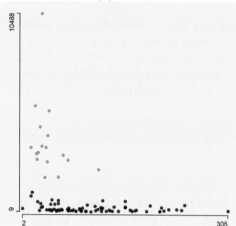

Age

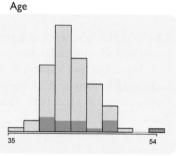

Age weighted by No. Taxpayers 1618

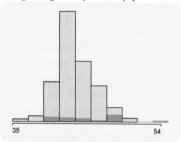

The tax districts with the highest tax payments are least populated. None of the highly populated districts contributes high tax payments.

High tax payments come almost exclusively out of tax districts with a high proportion of merchants.

There is neither a notable association between high tax payment and confession nor between high tax payment and age.

The most common average age is about 42 years. Weighting the average age of the districts with the number of underlying taxpayers shows the age distribution of the taxpayers.

High tax districts form a cluster.

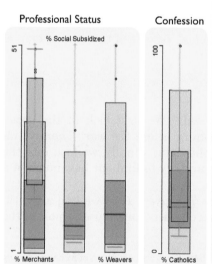

The scatterplots only show cases which are non-missing in both variables, i.e., for the three plots to the right, 18, 25 and 45.

% Merchants vs. % Weavers

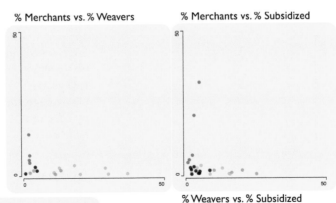

% Merchants vs. % Subsidized

% Merchants

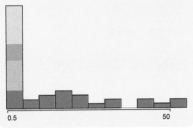

% Weavers vs. % Subsidized

% Weavers

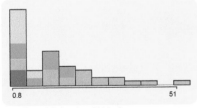

The three major professions form quite separate groups within the tax districts. Merchants separate strongest and can be found in a contiguous area in the center.

% Social Subsidized Workers

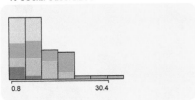

Merchants are only found in 45% of all quarters, whereas weavers were recorded in half and subsidized workers in two-thirds of all quarters.

% Merchants

% Weavers

% Social Subsidized

Tax 1618 vs. No. Taxpayers 1618

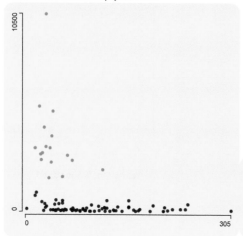

Tax payment and number of taxpayers on identical scales show no structural change between the two years.

On average, tax payments shrink by a factor of 4 and population by a factor of 2.5.

Tax 1646 vs. No. Taxpayers 1646

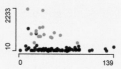

Tax 1646 vs. Tax 1618

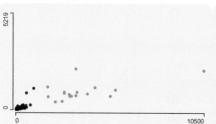

Tax 1646 vs. Tax 1618 (zoomed)

Tax payments in 1646 vs. 1618 (again on an identical scale) shows a decline of roughly a factor of 4. The zoomed view (right) shows a pattern similar to the overall pattern.

% Decline of Taxpayers

The percentage in taxpayers' decline is smallest in the wealthy city center (highlighting omitted).

No. Taxpayers 1646 vs. No. Taxpayers 1618

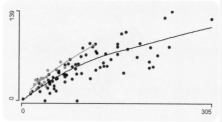

The decline in tax-paying population is stronger for the poorer and more populated districts.

tax payments. In contrast, all the highly populated districts contribute only a very small amount to the tax payments. Linking the wealthy districts to the histogram of the age distribution shows no particular pattern. To get a more faithful picture of the age distribution, we need to weight the histogram by the number of underlying taxpayers, which makes the distribution far narrower. In the parallel boxplot for the professions, we find that the wealthy districts are dominated by merchants. The highlighting in the map finally shows a contiguous cluster in the center of the city.

In the next step we want to classify the districts according to the major profession found in a district. The missing value plot shows that we have information on merchants for 45% of all districts, we have the share of weavers for half the districts and we have numbers for socially subsidized workers for two thirds of the districts. We need to keep this information in mind when looking at the three scatterplots and the three histograms of the proportions of professions. Districts where only one profession is found can directly be assigned. This works quite well for merchants who separate from the other two professions. The scatterplot of *% Weavers* vs. *% Subsidized* allows us to separate the two remaining groups. The color coding in the histograms nicely shows how much the professions mix or don't. The map reveals that the assigned groups form a geographically contiguous cluster.

The last step of the analysis looks into the change of tax payment from 1618 to 1646. The structure of the scatterplots of *Tax* vs. *No. of Taxpayers* does not change over the years. Whereas the payments shrink by roughly a factor of 4, the population shrinks by a factor of 2.5, but the L-shape remains the same. For a better comparison, the two scatterplots are on the same scale. The scatterplot of *Tax 1646* vs. *Tax 1618* shows an overall decline and is dominated by a few wealthy districts. Zooming in by almost a factor of 400 shows the self-similarity of the pattern for smaller tax payments. To assure the right interpretation, the aspect ratio of the two scatterplots has been set to one. The decline in tax-paying population is stronger for the poorer and more populated districts. The scatterplot smoother shows a larger slope, closer to one, for the selected, wealthier districts.

Further Analysis

A numerically more reliable estimate of the decline in tax payment can be generated by a linear model, which estimates the 1646 payments as a linear function of the 1618 payments.

```
> summary(lm(Tax.1646~Tax.1618-1,data=Augsburg))$coeff
          Estimate  Std.Error  t value      Pr(>|t|)
Tax.1618 0.2493374 0.01334434 18.68488 2.065732e-33
```

A differentiation regarding the major professions is left to the reader (cf. Exercise 4).

Summary

The analysis of the data shows a clear separation of the major professions at that time, especially of the merchants. The Thirty Years' War caused a major decline both in tax payments and population, but did not change the socio-economic structure of Augsburg.

For an explanation, we need to look at the historic background again. Due to the *Confessio Augusta* in 1530 and the *Augsburg Religious Peace* in 1555, Augsburg was a city of major interest for both sides. Although both opponents were interested in controlling the city, neither of them was interested in a destroyed city — Augsburg was too important a center of trade and supply. After the Swedes erected a new citywall, Augsburg was too well fortified to be attacked. The only kind of suppression against the city were attempts to starve it out.

Since there was no major structural change of the social and economic situation in Augsburg, the only remarkable tax districts are the two located outside the city wall. For these districts no taxpayers are recorded for the year 1646, although 90 resp. 38 have been found 28 years before. Although the data set does not reveal any information about these 128 taxpayers in particular, it leads to the policy: "Stay inside the city wall!" — which might not have been a good policy during the 1628 plague!

Exercises

1. Use Google Maps™ http://maps.google.com to locate Augsburg and compare the outline of the city in 1626 and the current inner city. How much has the city changed over the past 350 years?

2. Investigate the per capita tax payment decline from 1618 to 1646.

3. Calculate the relative tax decline and display the derived variable in a choropleth map. Look at the data also in a scatterplot of relative tax decline vs. tax 1618.

4. Extend the model from page 219 to take the dominant professions in the districts into account and compare the results to the initial model.

5. Discuss the advantages and drawbacks of coding the absence of a profession in a tax district via NA and not with a 0. Substitute all NA's with zeros for the three professions and use the function ternaryplot() from the R package vcd to plot the data. Do you get new or different insights?

H

Classification of Italian Olive Oils

Background

Olive oil, along with sunflower oil, soya oil and rapeseed oil, is one of the four most common edible oils. They account for more than 90% of all edible oils. Europe produces about 60% of the world's olive oil. Italy produces more than a third of this total, amounting to 28% of the world production only topped by Spain with more than 34%.

In order to assure world wide production quality and make products comparable, the "International Olive Oil Council" defines in Resolution NO. RES-3/89-IV/03:

2.1. **Olive oil** *is the oil obtained solely from the fruit of the olive tree (Olea europaea L.), to the exclusion of oils obtained using solvents or re-esterification processes and of any mixture with oils of other kinds. It is marketed in accordance with the following designations and definitions:*

2.1.1. **Virgin olive oils** *are the oils obtained from the fruit of the olive tree solely by mechanical or other physical means under conditions, particularly thermal conditions, that do not lead to alterations in the oil, and which have not undergone any treatment other than washing, decantation, centrifugation and filtration.*

2.1.1.1. **Virgin olive oils fit for consumption as they are** include:

(i) **Extra virgin olive oil:** *virgin olive oil which has a free acidity, expressed as oleic acid, of not more than 0.8 grams per 100 grams, and the other characteristics of which correspond to those fixed for this category in this standard.*

(ii) **Virgin olive oil:** *virgin olive oil which has a free acidity, expressed as oleic acid, of not more than 2 grams per 100 grams and the other characteristics of which correspond to those fixed for this category in this standard.*

(iii) **Ordinary virgin olive oil:** *virgin olive oil which has a free acidity, expressed as oleic acid, of not more than 3.3 grams per 100 grams and the other characteristics of which correspond to those fixed for this category in this standard.*

Parallel to this definition, the USDA defines grades A to D in the "United States Standards for Grades of Olive Oil" from March 22, 1948 in §52.1532 as not having more than 1.4% oleic content (A), less than 2.5% (B), less than 3% (C) or more than 3% (D). Obviously the two scales do not match well and the requirements for extra virgin olive oils are not necessarily met by grade A oils.

No matter what exact thresholds are used for the definitions, both scales use the content of the fatty acid *Oleic* as a measure of quality. From all fatty acids found in olive oils, the primary fatty acid is *Oleic* with an average 66%. Other fatty acids are *Linoleic* 12%, *Palmitic* 9%, *Eicosenoic* 5% and *Palmitoleic* 5% plus traces of others.

A classification of different kinds of olive oils from different countries, areas or regions can thus be made by examining the composition of the various fatty acids in the olive oils.

Goals of Study

The general goal of the analysis is to find rules that allow us to distinguish olive oils from different regions via their fatty acid content. Although this task is a classical classification job that can be done more efficiently with classification methods such as decision trees or machine learning techniques, the graphical analysis can give more structural information on the dataset. Graphical tools can diagnose what is usually hidden in the black box of a classification algorithm.

Description of Data

The data used in this case study are the 572 Italian olive oils from the paper of Forina et al. (1983). Their study used samples from Italy, Greece, Crete, Lebanon, Syria, Israel and Portugal. For the Italian sample, eight fatty acids have been recorded along the *Region* where the sample was taken. The nine *Regions* can be hierarchically grouped into three *Areas*.

The actual dataset taken for the analysis can be found at the website of Zupan and Gasteiger (1999) which covers the dataset in Chapter 10 (pp. 176–189). All variables have been scaled to fit a range of $[0, 100]$. Such a normalization is usually not necessary when working with graphics, but will turn out to be quite handy when we apply classification methods to the data, which will be used to compare with the graphical analysis results.

Figure H.1 shows a map of Italy which highlights the nine regions from which the 572 olive oil samples were taken.

FIGURE H.1
A map of Italy highlighting the nine regions.

Graphical Analysis

The first step in a graphical analysis of such data is to use overview plots such as a scatterplot matrix (SPLOM) or a parallel coordinate plot for the eight continuous variables and barcharts for the two categorical variables. Overplotting is an issue for both the scatterplot matrix as well as the parallel coordinate plot when we use color brushing instead of highlighting. The use of color brushing can visualize the hierarchic relationship between the variables *Area* and *Region*.

Picking the most interesting features from the scatterplot matrix and the parallel coordinate plot, we start to look at possible artifacts in the data. All scatterplots of *eicosenoic* show that all oils from the *North* or

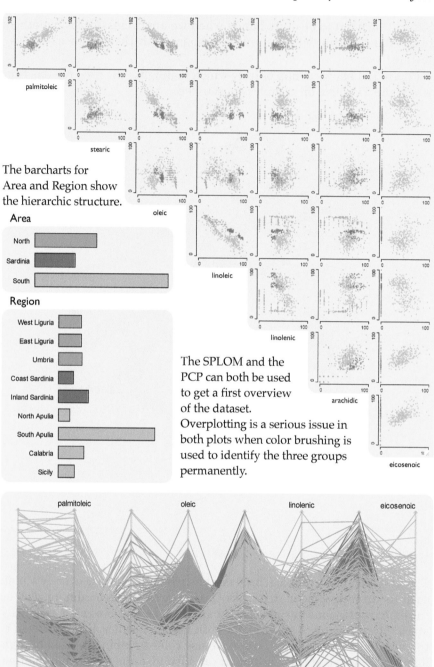

The barcharts for
Area and Region show
the hierarchic structure.

Area

North

Sardinia

South

Region

West Liguria

East Liguria

Umbria

Coast Sardinia

Inland Sardinia

North Apulia

South Apulia

Calabria

Sicily

The SPLOM and the
PCP can both be used
to get a first overview
of the dataset.
Overplotting is a serious issue in
both plots when color brushing is
used to identify the three groups
permanently.

Area

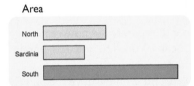

Selecting all southern oils shows that they are the only group that is not all 0 for the eicosenoic oil content.

Linoleic vs. Eicosenoic

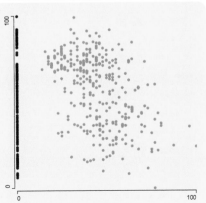

There is a strong discretization in the variables arachidic and linolenic, and a still noticeable rounding effect in stearic and pamitoleic only for the subgroup of Ligurian oils. Setting the histograms to small binwidths or increasing the α-transparency for non-highlighted points reveals the effect which can be attributed to traceability limits.

Region

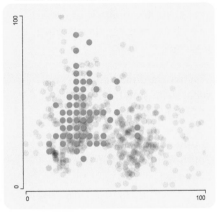

Arachidic

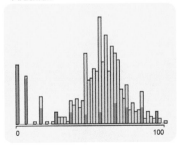

Arachidic vs. Linolenic

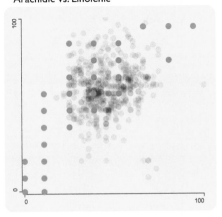

Stearic vs. Palmitoleic

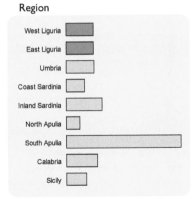

Region

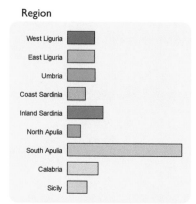

When trying to separate the nine regions, scatterplots can be used to find axis-parallel splitting rules.

Because Oleic and Linoleic have the biggest shares of all fatty acids they can provide the most information.

Eicosenoic separated southern oils from the two other groups and adding Linoleic can separate the two Sardinian oils from the northern oils.

Linoleic vs. Oleic

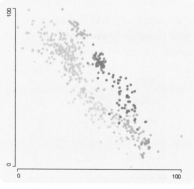

Linoleic vs. Eicosenoic

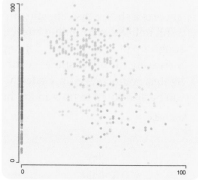

Once the three areas can be distinguished from each other, parallel coordinate plots can be used to look for further rules.

For the northern oils, it can be seen that Umbrian oils fall into a quite narrow band.

The two Sardinian oils can be separated using the variables Oleic and Linoleic.

Southern oils are dominated by South Apulia. Sicilian oils are dispersed.

North

Sardinia

Sardinia have an eicosenoic acid content of zero. In particular, the scatterplot of *linoleic* vs. *eicosenoic* shows a separation of all three areas.

Both the scatterplot matrix and the parallel coordinate plot show a discrete pattern for four variables, namely, *Linolenic, Palmitoleic, Arachidic* and *Stearic*. Selecting the Ligurian olive oils shows that this feature is restricted to this region. The discretization can be visualized in a histogram, for instance, for *Arachidic* by using a small binwidth of 5%. The feature also shows up strikingly when observing any pair of the four variables in a scatterplot with increased α-transparency for all non-highlighted points.

All cases can be color-brushed by *Region* to find splitting rules which discriminate among all nine regions. The scatterplots of *Linoleic* vs. *Oleic* and *Linoleic* vs. *Eicosenoic* offer many potential splits to separate regions. Using the subsamples for northern, Sardinian and southern oils, we can use parallel coordinate plots to look for further possible splits to separate the regions in the dataset. For consistency, the color brushing remains the same for all plots.

Further Analysis

Forina et al. (1983) use eight different methods, ranging from eigenvector projection over (modified) linear discriminant analysis to k-nearest neighbor methods. The technique we want to investigate more closely here is decision trees. Assuming the dataset to reside in the R-dataframe called olives, we can construct and test a classification tree by:

```
> # load the rpart-library
> library(rpart)
> # create the tree model (exclude Area!)
> t1 <- rpart(Region ~ . , data = olives[,1:9])
> # create confusion matrix
> t2 <- table(predict(t1, type="class"), olives$Region)
> # everything not on the diagonal is an error
> sum(t2) - sum(diag(t2))
[1] 42
```

The resulting tree is visualized in Figure H.2 using the interactive software KLIMT (see Urbanek 2006 for further details).

Keeping in mind that a set-up without independent training and test data will usually underestimate the classification error, the resulting tree has 42 misclassified cases. We can then use a fluctuation diagram to visualize the confusion matrix, i.e., the table of true class labels vs. predicted class labels (Figure H.3 left). We find that most errors are within the same area and only a few can be found across areas. Sicilian oils are clearly the ones most frequently misclassified.

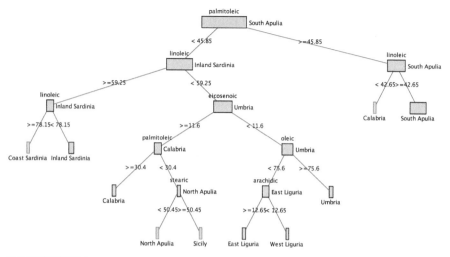

FIGURE H.2

A binary classification tree for *Region* generated with the R-function
rpart() and visualized in KLIMT.

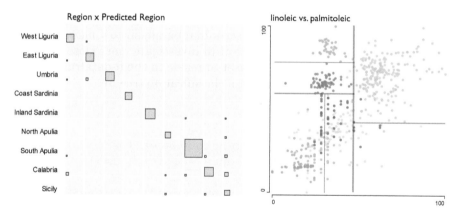

FIGURE H.3

Left: A fluctuation diagram to visualize the confusion matrix (true label
in the columns). Right: A sectioned scatterplot that shows five out of the
nine splits in the tree of Figure H.2.

As five out of the nine splits in the tree in Figure H.2 can be found in
the scatterplot of *Linoleic* vs. *Palmitoleic*, we can use this plot to visualize
most of the splitting structure of the tree. Such a scatterplot is called a
sectioned scatterplot. Four out of the six regions in Figure H.3 right refer
directly to terminal nodes in the tree in Figure H.2.

Summary

From all plots that depict *Eicosenoic* acid, we can see that this acid was either not found or at least too small to be quantified in all regions except *South*. Forina et al. (1983) state: *"...Data reported as 'trace' in the original literature were processed as a random number in the range between 0 and 0.1%. ...* (page 193)" The data from Zupan and Gasteiger (1999) obviously contain the "true" zeros.

The discretization found for four of the variables for the Ligurian oils can most probably be attributed to a limited resolution of measuring the fatty acid content with different standards and machinery. Methods and algorithms applied to this data may accidentally pick this artifact as a signal and produce misleading results.

When trying to classify the data graphically, we see that the three areas can be separated quite easily. The Sardinian oils can be split without any error as well. Umbrian oils can be distinguished from Ligurian oils, which are only slightly intermixed. The southern oils are most problematic. Once the oils from *South Apulia* are identified, Sicilian oils are scattered in all dimensions and make up most of the classification error. The hierarchical approach to first find rules for the areas and to then classify the remaining data in the three subgroups is quite effective. Graphical methods are only of limited use to classify datasets in a high-dimensional continuous space. Methods such as decision trees, model-based clustering or support vector machines are better at this task.

The strength of graphical methods can be found in the diagnostics of classification methods. All plots that rely on an axis-parallel visualization of the data are ideal for visualization of the mechanics of a classification tree. Methods such as support vector machines constructing almost arbitrary decision boundaries need more sophisticated visualization techniques based on multivariate scatterplots , e.g., as provided by ggobi (see Cook et al., 2004). Nonetheless, the interpretation of a decision boundary in more than three dimensions remains a non-trivial task.

Exercises

1. Construct a decision tree for the three areas. Force the decision tree for the nine regions to have the same root split as the tree for the three areas. Does the number of misclassified cases improve?
   (Hint: the tree generation algorithm can be run independently on the two branches of the desired tree.)

2. Take a sample of 2/3 of the original size of the dataset, respecting the class proportions. Use this sample as a training dataset to construct a classifier. Use the remaining data to validate the classifier. How does this change the results, and how much do they depend on the actual sample chosen?

I

E-Voting in the 2004 Florida Election

Background

The United States pres-
idential election in the
year 2000 attracted
global attention for be-
ing extremely close and
controversial. The state
of Florida become the
center of controversy due
to its decisive role as only
537 votes out of a total of
5,963,110 separated the
presidential candidates.

Although all regis-
tered voters can cast
their votes, the president
of the United States is
not elected by popular
vote. Instead, each state
chooses United States Electoral College electors who represent the
state and cast their votes. Traditionally, the electors cast their votes in
accordance with the popular vote in their states. The number of electors
for each state is determined approximately by the population of each
state;* thus a populous state such as Florida can easily be pivotal.

The fact that a very small number of votes played such an important
role also highlighted several problems in the voting process. One such
issue was the way votes were cast, including using punch cards that left
room for error. Electronic voting machines were used for the subsequent
2004 presidential election as a possible remedy for the problem. However, the implementation of electronic voting itself raised many questions.
The voting machines did not produce any paper trail, leaving room for

---

*More precisely the number of electors is equal to size of the United States Congress (since
1961 consisting of 435 Representatives + 100 Senators) plus electors allocated to Washington, D.C. (3). The House of Representatives is apportioned by population.

speculations about manipulations. Those were fueled by the closeness of some manufacturers to one of the candidates. In addition, several vulnerabilities were discovered before the election. Consequently, the final results came under scrutiny to see whether there were indications that electronic voting favored one or the other candidate. An analysis by Hout et al. (2004) that challenged George W. Bush's win in Florida has become widely cited.

Goals of Study

In the following we want to analyze the 1996–2004 United States presidential election results for the state of Florida summarized by county, with special focus on electronic voting. The basic idea of many model-driven analyses is to eliminate possible effects and see whether the remaining variance in the data can be explained by the electronic voting effect. Therefore we add covariates as proposed by Hout et al. (2004) to assess their influence on the voting outcome.

Description of Data

The data used in this study are based on the sources used by Hout et al. (2004) to allow direct comparison.[†] The data for years 1996 and 2000 are based on official voting results per Florida county; data for 2004 were taken from the latest CNN report. Demographic variables are based on the 2000 census and electronic voting type as reported by Verified Voting Foundation.

County - county name

FIPS - county FIPS (identification) code

Dole 1996 - votes for Bob Dole (R) in 1996 election

Clinton 1996 - votes for Bill Clinton (D) in 1996 election

Bush 2000 - votes for George W. Bush (R) in 2000 election

Gore 2000 - votes for Al Gore (D) in 2000 election

Bush 2004 - votes for George W. Bush (R) in 2004 election

Kerry 2004 - votes for John Kerry (D) in 2004 election

e-Voting - electronic voting indicator variable (1=yes, 0=no)

Median Income - median household income in the county

Hispanic Population - proportion of Hispanic population in the county

---

[†]We also provide a more complete dataset based on official sources. It is available from the book's website.

In the course of analysis we use several derived variables, such as *Votes* per year consisting of the sum of votes for both candidates, relative support (votes for a candidate divided by *Votes* for the year — denoted by appending % to the name) and *Swing* which is the difference of relative support for a candidate between two consecutive election years (named after the later year).

Graphical Analysis

The dataset is organized by county which allows us to visualize all features both in a map and in conventional plots. Using multiple map views has the benefit of direct comparison of features as long as the color mapping is chosen reasonably. The number of votes cast (e.g., *Votes 2000*) has to suffice as proxy for county population. Comparisons of votes should be done using proportional measures in order to eliminate the effect of large population differences. We look at the *Bush 2000 %* relative support as a baseline to study which factors may be related to it. Since our data are restricted to votes for Democratic and Republican candidates only, it is sensible to choose a symmetric interval centered at 50% for display. To follow the convention we use a blue-white-red color scheme to indicate high Republican support as red and high Democratic support as blue. Comparing Republican support to other factors, neither proportion of *Hispanic Population* nor the *Median Income* per county seem to match the pattern. Only higher county population appears to be linked to higher Democratic support.

The question at stake is whether electronic voting influenced the election. One way to look at it is to consider the 'swing,' i.e., the difference in the relative support for Bush between the years 2004 and 2000. This measure takes out the effects of population fluctuation, turnout rate change and global candidate support level. Any significant difference caused by e-voting should show in this measure (unless cancelled by an equal effect), because e-voting was not used before the year 2004. A look at *Bush Swing 2004* shows that most counties except for a very few increased their support for Bush, including the largest county, *Miami-Dade*, which also used e-voting. In general, no direct effect of e-voting is visible.

We have seen earlier that the size of the county may be correlated with Republican support. To analyze this more closely, we plot *Bush 2000 %* vs *Votes 2000*. Again, we use year 2000 as a baseline. Adding a loess smoother illustrates the global trend of decreasing Bush support with increasing county population. Note that by adding colors corresponding to e-voting, we see that all of the most populous counties used e-voting.

Since the e-voting variable is dichotomous, spinograms are very useful in assessing its effect on other continuous variables, especially with density estimates. However, keep in mind that we can only detect patterns, not causality. An example is *Bush 2004* variable that shows a very clear

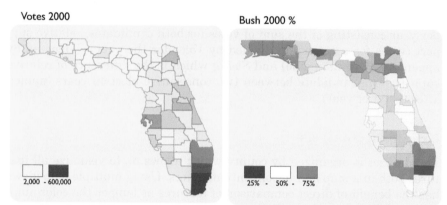

Number of votes as a proxy for county population appears to be reflected in the Democratic support for most large counties.

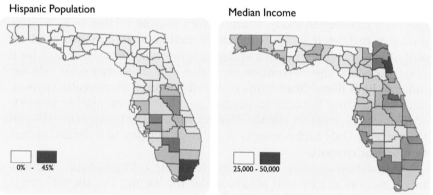

Neither demographic factor shows a clear relationship to election results on its own. Corresponding scatterplots confirm this.

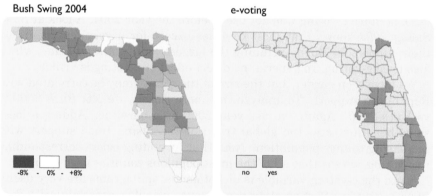

The support for Bush increased from 2000 to 2004 in most counties. However, no clear pattern is visible with respect to electronic voting.

Bush 2000 % vs Votes 2000

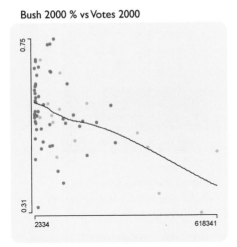

Relative support for Bush decreases with the size of the county, represented here by the number of votes.

e-Voting

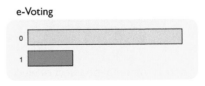

Bush 2004

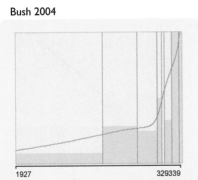

Votes 2004

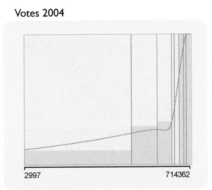

The effect of electronic voting on absolute support is apparent yet unchanged between 2000 and 2004. It corresponds to the relationship of e-voting and county size.

Bush 2000 %

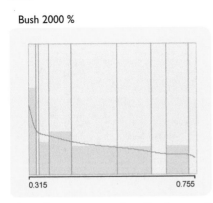

Bush 2004 %

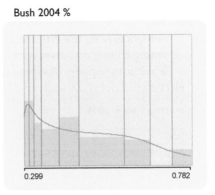

The relative support hardly changes between 2000 and 2004 with respect to counties that introduced e-voting in 2004.

Bush Swing 2000 | Votes 2000

Bush Swing 2004 | Votes 2004

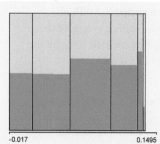

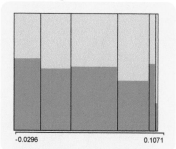

There is virtually no effect of electronic voting on the votes by swing. Interestingly, the pattern has even shifted slightly as if the e-voting favored the Democratic candidate in 2004 compared to the previous election.

Bush 2004 % vs Bush 2000 %

Bush 2000 % vs Dole 1996 %

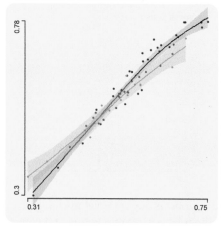

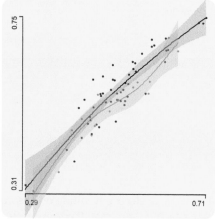

Bush Swing 2004 vs Bush 2000 %

Broward and Palm Beach counties are outliers driving the difference for e-voting among countries with low Republican support. The previous election did not show such an effect.

e-Voting | Votes 2004

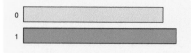

The outlier status is better visible when plotting swing vs support.

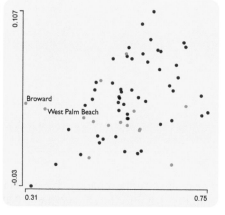

pattern of increasing e-voting proportion with increasing number of votes for Bush. However, the reason is not e-voting because the same pattern can be seen in earlier years without e-voting and in the total number of *Votes 2004*. The latter is the reason, because large counties have introduced e-voting. Relative support takes out this effect and shows a very slight decrease in e-voting along relative support. However, a comparison with the year 2000 election shows a very similar pattern and thus is not conclusive.

We have mentioned that swing may be a good measure since it adjusts for several effects. However, comparing e-voting proportions in swing directly is not fair as it will count each county equally regardless of its size. Thus the areas in the spinogram will correspond to the number of counties and not votes. In order to account for the number of votes supplied by each county, we use a weighted spinogram of *Bush Swing 2004* with *Votes 2004* as weights. Although the *Bush Swing 2004* is still partitioned the same way as in a regular spinogram, the contributions within bars are now scaled by county size. Thus all areas in the spinogram represent votes. We show such a weighted spinogram both for 2000 and 2004 elections.

The first observation is that about half the votes were cast by e-voting in 2004. This is true and can be verified by plotting a weighted barchart of *e-Voting* with *Votes 2004* as weight. The second observation is that the proportion of e-voting across swing is nearly constant in both years. Interestingly, the 2000 election shows a very faint trend of increasing proportions with Republican support, whereas 2004 shows a very faint trend in the opposite direction, in contrary to the allegations.

We could not see any clear pattern across swing, so let us have a look at *Bush 2004 %* vs *Bush 2000 %* directly. At a first glance we see a clearly linear relationship as expected but no obvious outliers. Adding smoothers for each group — e-voting and paper voting — allows us to compare each relationship separately. We see an intersecting trend: counties with lower Republican support and e-voting supported Bush more, whereas counties with strong Republican support and e-voting supported him less. *Bush 2000 %* vs *Dole 1996 %* serves as a sanity check and this time differs, in particular in the region of low Republican support. We can see that two counties drive the rise in the year 2004 plot, thus we can use *Bush Swing 2004* vs *Bush 2000 %* for confirmation. *Broward* and *West Palm Beach* are those two counties.

Further Analysis

At this point we want to demystify whatHout et al. call "excess votes in favor of Bush associated with Electronic Voting" and estimate at 130,733 votes. It is the difference of the predictions from their model by using e-voting variable (prediction 1) and by setting it to zero for all counties (prediction 2). This number can be recalculated with the following R-code:

```
> # rebuild Hout's model
> hm <- lm(bush_change ~ bush2000pc + bush2000pc_sq +
                 votes2004 + evote + bush2000pc_evote +
                 bush2000pcsq_evote + votes_change +
                 dole1996pc + income + hispanic, data=e)

> p_votes <- (predict(hm) + e$bush2000pc) * e$votes2004

> # create data with e-voting "removed"
> e0 <- e
> null <- rep(0, length(predict(hm)))
> e0$evote <- null
> e0$bush2000pc_evote <- null
> e0$bush2000pcsq_evote <- null

> # predict the new data with Hout's model
> p0 <- predict(hm, newdata=e0)
> p0_votes <- (p0 + e$bush2000pc) * e$votes2004

> # calculate the difference
> sum(p_votes-p0_votes)
130733.8
```

It is an easy task (cf. Exercise 2) to investigate how unreliable and unstable this figure is, based on the over-parametrized model.

Results

The support of a Democratic contender increases with the population of a county, but that was the only conclusive single effect that could be distinguished. Although other spatial patterns became apparent, they did not influence the voting behavior substantially or systematically.

Electronic voting was introduced by all five of the most populous counties which also contributes to the fact that the majority of votes was cast electronically. The effect of electronic voting on the results is not obvious, although two counties, *Broward* and *West Palm Beach*, show considerably high swing in favor of Bush in 2004 despite their overall high Democratic support. Those two counties were driving the models of Hout et al. and thus making the generalization of their results questionable. The fact that those highly populous counties account for most noticeable differences cannot silence speculation about voting machine overflows and other issues that were raised.

Nonetheless the analysis illustrates the perils of jumping to conclusions and reinforces the necessity to verify all findings by using a different view or comparison to similar data — in this case from the previous elections.

Exercises

1. We have used choropleth maps to visualize individual covariates and look for simple relationships. What other plots could be used for the same purpose? What are their advantages and disadvantages compared to maps?

2. Martin and Broward counties have the highest leverage in Hout et al.'s model. Recalculate the number of "excess votes" with these two cases removed from the data.
   How large is the figure when you use the simpler Model 2?

3. The dataset analyzed here is not based on the official results. Download the complete dataset with official, final results from the book webpage.

   (a) Perform the same analysis with the official data. Has anything changed?

   (b) The official data also contain votes for other candidates. What influence did their support have in 2000 and 2004 on the Republican or Democratic candidate in each of those years? Can you see any outliers?

i

Mondrian Reference

This Appendix gives a brief overview of the Mondrian software. It covers the most important plots and commands. More information can be found at the Mondrian websites. The underlying concepts that need to be known in order to work efficiently with Mondrian are described in the first part of this book and are not mentioned in this reference.

i.1 Quick Start Guide

Installation and Startup

Mondrian can be downloaded at either www.rosuda.org/Mondrian or www.theusRus.de/Mondrian. For Windows and MacOS X, Mondrian comes as a native application. Whereas on MacOS X a suitable JAVA Virtual Machine will be pre-installed, Windows users may have to install a recent Java Runtime Environment (JRE) from Sun Microsystems if it is not already installed on their machine. To start Mondrian simply double-click the application icon; no further installation is necessary. LINUX users need to download the jar-file of Mondrian, and start it manually with the command java -jar Mondrian.jar.

Loading a Dataset

Mondrian reads tab-separated or space-separated text files that feature a header-line for the variable names. It reads alpha-numerical data as well as numerical data and supports scientific notation. Missing values must be coded as "NA."

When exporting data from Microsoft Excel, choose the *Text*-option. Note, depending on your local language settings, Excel might use a "," as the decimal point, i.e., non-scientific notation. If this is the case, you need a text editor to replace "," with ".".

R users can export a data-frame with the simple command write.table(Data, "MyFileName.txt", quote=F, sep="\t", row.names=F, header =T) to make it readable in Mondrian.

To load a dataset, simply choose File > Open and select the desired file. If there is an irresolvable error while reading the data, Mondrian will show an error message that points to the problem as precisely as possible.

The Variable Window

The variable window is at the core of any Mondrian session. It shows an entry for all variables read from the datafile. When reading a datafile, Mondrian automatically determines the scale of each variable, which is then displayed via a little icon in front of each variable. This can be

abc alpha-numerical, i.e., a categorical variable containing text.
     (The scale of an alpha-numerical variable cannot be changed.)

 discrete, i.e., a numerical variable with only a few distinct categories.
     (The scale can be changed to continuous.)

 continuous, i.e., a numerical variable with many different distinct categories compared to the overall number of cases.
     (The scale can be changed to discrete.)

Whenever a variable contains missing values, the background of the corresponding icon turns white.

Sometimes, a very coarse resolution makes a continuous variable look like it is discrete, or a discrete variable is displayed as being continuous due to many levels and only relatively few observations in the dataset. In these cases, a variable type can be changed by a simple double click on the variable's name. If there is a whole group that needs to be changed, it is easier to select all the variables and use the shortcut <meta>-t.

To make a change permanent, the variable names in the datafile can be modified with the prefix /D to force an interpretation as discrete variable or /C to force an interpretation as continuous variable. The variable window also shows the name of the dataset in the title bar, and the number of cases in total and the number of selected cases in the status bar.

Throughout this Appendix, we often use the *Cars 2004* dataset, which comes from the data archive of the *Journal of Statistical Education* http://www.amstat.org/publications/jse/jse_data_archive. html, and which can be downloaded at the book's website. In Figure i.1 we find that *City Miles per Gallon* and *Width (inches)* are both marked to be discrete. As the width is measured in full inches only, it ranges from 64 to 81 and thus takes only 18 distinct values, which is relatively few for 428 cases in the dataset.

The order in which variables are selected is respected in all plots that show more than one variable, e.g., scatterplots or all multivariate plots.

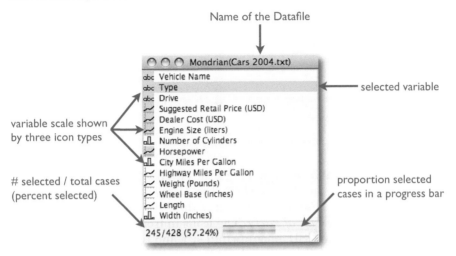

Name of the Datafile

selected variable

variable scale shown by three icon types

selected / total cases (percent selected)

proportion selected cases in a progress bar

FIGURE i.1
The variable window for the ***Cars 2004*** dataset. 245 out of 428 cases are selected, making up 57.24% of the sample. The alpha-numerical variable ***Type*** is selected. About half of the variables contain missing values.

A Hands-On Case Study: The Tips Data

In this subsection we run through a small case study as a hands-on, step by step example to illustrate some of Mondrian's functionality. The data are taken from case study 18 in Bryant and Smith (1995). A food server recorded data for about two and a half months in early 1990. The restaurant was located in a suburban shopping mall and part of a national chain of restaurants. There was a smoker and a non-smoker section in the restaurant.

The idea behind collecting such data is to find factors that might influence the tipping behavior. Analyzing tipping behavior has been of interest for a long time and several other studies have been carried out. For instance, Azar (2002) talks of an estimated amount of 26 billion USD in tips per year in the U.S. — just in restaurants. Conlin et al. (2003) performed a similar study and noted data on over 1,000 customers with even more detailed data. Whereas most of the studies try to build statistical models to describe the tipping behavior and find statistically significant factors, we will focus more on the exploratory analysis of such data. In this introductory section we use the 244 observations of the simple dataset from Bryant and Smith (1995).

We start the analysis with loading the data. In the menu, we can choose > File > Open to locate and load the tips dataset:

The resulting variable window shows the seven variables in the dataset. The *Bill in USD*, the *Tip in USD*, the *Gender* of the paying person, whether there was a *Smoker* in the party, the *Weekday*, the time of the day *Day/Night* and the *Size of Party*. The icons in front of the variables indicate their scale. The USD amounts (bill size and tip) are continuous, *Size of Party* is numerical discrete, and all other variables are alpha-numerical discrete.

The first step in an analysis is to get an overview of the data and create low-dimensional plots for all variables. Select *Bill in USD* and by holding the <ctrl>-key (Win) or the <command>-key (Mac) add *Tip in USD*.

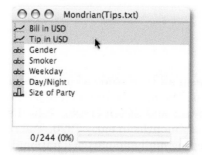

Then go to the Plot-menu and choose > Plot > Histogram. The following two histograms will show up in default parametrization.

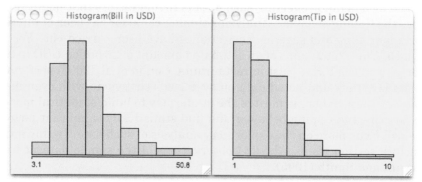

We can create barcharts for the remaining five categorical variables.

Select the variables in the variable window and choose > Plots > Bar-chart in the Plot-menu.

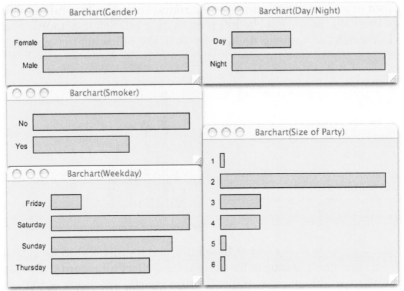

From the barcharts we see that more than two-thirds of the data were recorded during the day and 2/3 of the bills were paid by male customers. About 40% of the cases were recorded in the smoking section. Very few cases can be found on Fridays. The vast majority of parties were of size two and only rarely are parties of size one or five and above noted.

The scale of the histogram shows that bills ranged from just above $3 up to almost $51. Tips range from exactly $1 to $10. From querying (<ctrl>mouse-over) the tallest bar in the histogram of *Bill in USD* we learn that most bills are between $14 and $19.

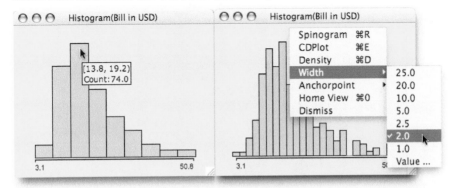

To get a better interpretation, we set the anchorpoint of the histogram to $2 and the binwidth to $2 using the context menu (right click or <ctrl>click on Mac). To re-fit the histogram press <ctrl>-0 (Win) or

<command>-0 (Mac) or use home in the context menu. The histogram for *Tip in USD* has only nine bins in the default view. We know from our own experience that many people round their tips to a full amount or align their tips such that the bill plus tip is a full dollar amount. Setting the bin width to a quarter dollar results in the following histogram.

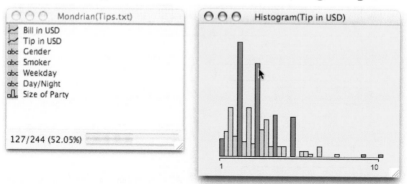

Selecting all bars that cover full dollar amounts (use <shift>-click to select multiple bars) accounts for more than 50% of all records. The status bar in the variable window indicates that exactly 127 cases out of a total of 244 cases (52.05%) are selected.

In the U.S., there is a rule of thumb that the tip should not be less than 15% of the bill. To investigate the relationship between bill size and tip, we can plot the scatterplot of *Tip in USD* vs. *Bill in USD*. The variable that is selected first is assigned to the *x*-axis, the second to the *y*-axis.

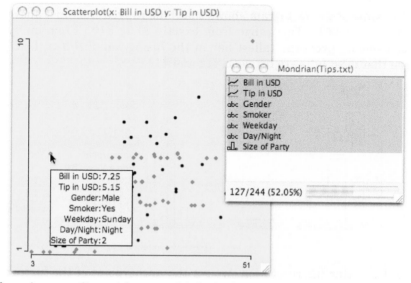

There is an outlier with a very high tip for a rather small bill. Querying this point we find a bill size of $7.25 and a tip of $5.15. To query all data related to this point, we can select all variables in the variable window and use an extended query (<ctrl>-<shift>-mouse-over) for the outlier.

For a further analysis of the tipping rate, i.e., tip divided by bill, we need to derive a new variable. Selecting *Bill in USD* as nominator first and *Tip in USD* as denominator second, we can go to the Calc-menu and choose > Calc > transform > Tip in USD / Bill in USD to calculate the tip rate. Note that the variable names are automatically inserted in the menu for convenience.

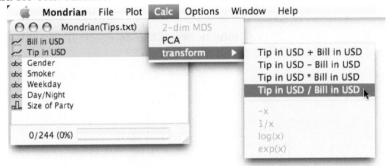

Using a very fine resolution in the histogram of *Tip in USD* / *Bill in USD* of 1%, we can identify two peaks at 15% and at 19%.

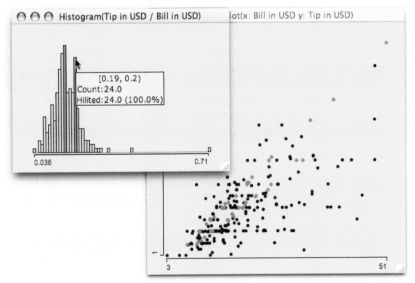

Selecting these two peaks shows the linear relationship in the scatterplots. From the selection we can conclude that 20% is a threshold that is only rarely exceeded when tips are given.

Before we look at the relationship between the categorical variables, we want to bring the weekdays into their natural order. Alt-dragging the bar of Thursday — which ends up being the last in lexicographical order

— to the first position gives the desired order.

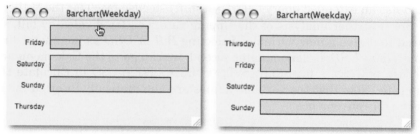

We look next at two interactions between the categorical variables *Gender* and *Smoker*, and *Weekday* and *Day/Night*.

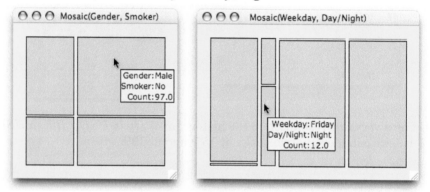

In an interactive environment we use queries instead of lengthy labels. Smoking turns out to be independent of the gender. *Weekday* and *Day/Night* are very unbalanced. There is no information during the day on Saturdays and Sundays, and almost no data for Thursday nights. As long as we do not look at more than 2 dimensions, we might as well use linked barcharts and spineplots to get the same kind of information.

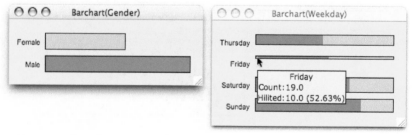

Creating two barcharts, one for *Weekday* and one for *Gender*, selecting males and switching the barchart of *Weekday* to the spineplot view via <ctrl>-r (Win) or <command>-r (Mac) reveals that the proportion of males paying the bill grows steadily from less than 50% on Thursdays to more than 75% on Sundays.

After we have a fairly good understanding of all variables and their basic interactions, we finally want to investigate which factors do influence the tip rate. First, we look at a possible functional relationship between *Bill in USD* and *Tip in USD*.

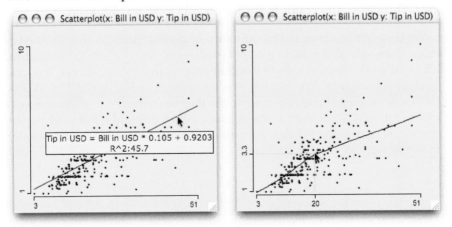

In the context menu of the scatterplot *Tip in USD* vs. *Bill in USD* choose > smoothers > ls-line for a linear fit. Querying the regression line shows an offset of one dollar and an increase of 10.5 cents per dollar of bill size. This function definitely does not fit the data well for small bill sizes so we switch the smoother to a loess smoother. With shift-up/down arrow we can change the smoothness interactively and look for a suitable fit. A span of 0.75 shows two almost linear pieces, one up to $20 and one above $20, which seem to fit the data better. An overview query (alt mouse-over) makes it easy to identify the cut-off at 20 dollars.

Selecting *Tip in USD / Bill in USD* and *Size of Party*, we can choose > Plots > Boxplot y by x to compare the tip rate for different party sizes.

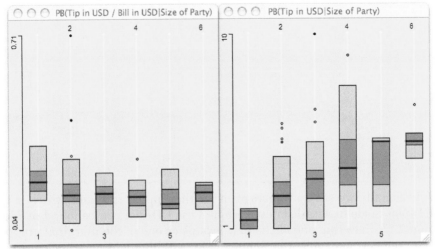

The tip rate declines as party sizes increase. On the other hand, the absolute tip obviously increases with party size as bills will be larger when more people order. Checking the influence of *Weekday* on the tip rate gives the surprising result that there seems to be no influence.

What is left to check is the influence of *Gender* and *Smoker* on the tip rate. Using the the the mosaic plot of *Gender* and *Smoker* set to fluctuation diagram mode, we are able to look not only at individual effects, but also at their interaction. We also use the scatterplot of *Tip in USD* vs. *Bill in USD* with a smoothing spline with one degree of freedom which is equivalent to a least square fit, but gives us confidence bands. For judging the interactions, we display the tip rate conditioned on *Smoker* and *Gender* in boxplots.

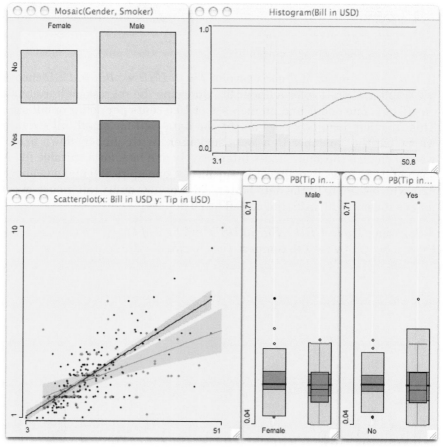

Additionally we set up a histogram for the tipping rate and switch it to CD-plot mode (a detailed description of the CD-Plot can be found in Section 2.2). Selecting one of the four groups or one of the marginal groups (i.e., smoker/non-smoker or female/male) will highlight possible features.

There is no relevant difference in the medians of the boxplots for tip rate when split by *Gender* or by *Smoker*. Selecting individual boxes in the fluctuation diagram, reveals a relatively strong effect for smoker. Female smokers tend to have an above average tip rate with relatively small variance. Male smokers in turn are more associated with smaller tip rates and have greater variation. In the CD-plot we find that the group of male smokers is more often found with higher bills, and female smokers more often with smaller bills.

Summing up, we found that the usual tipping rates range from 15% to 20%. Higher tipping rates are found with extremely small bills only, smaller rates are usually associated with high bills. The co-variables *Day/Night*, *Weekday*, *Gender*, *Smoker* and *Size of Party* seem to have only an indirect influence on the tip rate via the bill size, which in turn is strongly dependent on the party size. *Day/Night* interacts strongly with *Weekday*, and party sizes of one, five and six were rarely observed, which makes it difficult to draw valid conclusions from these very small subsets of data.

This hands-on case study showed only the most important aspects of the data. More graphs can be looked at and further conclusions might be drawn. As in the case studies in the second part of the book, the cases studies address specific questions. Other questions and viewpoints might lead to other graphics and aspects of the data. The case studies in Part II also show only graphs that are results of an intensive exploration of the data. No in-between steps – often of a technical nature — are shown as in this case study. The most important point is to have flexible and interactive tools, which support an exploratory work style.

i.2 Plots

The previous section used several of Mondrian's plot types and their options. This section introduces all plot-types in Mondrian and their plot specific commands which can either be triggered by the plot specific context menu or by keyboard shortcuts.

Missing Value Plot

The Missing Value Plot is only enabled if there are missing values in the data at all. Variables with no missing values will be ignored in this plot. If the user creates a missing value plot for variables that all have no missing values, a warning will be shown.

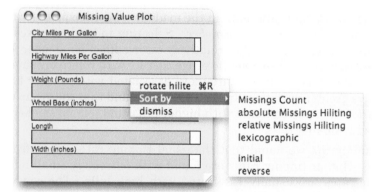

The missing value plot features sorting options similar to the barchart. Additionally, the initial order of the variables can be restored. The highlighting can be rotated to make the highlighting proportions between observed and missing values comparable.

Barchart and Spineplot

Barcharts and Spineplot live in the same plot window. A barchart can be switched to a spineplot and vice versa.

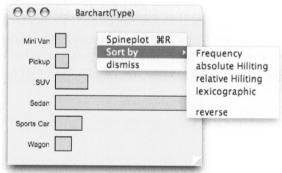

The categories in a barchart can be sorted manually by <alt>-dragging the bar to its new position. If dropped on a bar, it will exchange with that bar; if dropped between two bars, it will be inserted. The categories can be automatically sorted according to name, size, absolute and relative highlighted and reversed. If a single category is completely highlighted, <page-up> and <page-down> selects the previous or next category.

Histogram and Spinogram

Histograms and Spinograms feature the same duality as barcharts and spineplots.

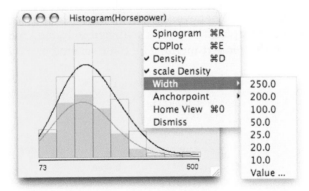

Histograms can be switched to spinograms and vice versa. The anchor-point and bin width can either be set to one of the pre-calculated "even" values or set to an arbitrary value. Both parameters can also be changed by keyboard controls. ← and → will move the anchorpoint by 1/10th of the current bin width to the left or right. ↑ and ↓ will increase and decrease the bin width by 10% of the current bin width.

The home-view command fits the histogram to the plot region after a change of the parameter. Zooming constrains the minimum in y to be 0.

Histograms can show an added density estimator calculated by the R-function `density`. The bandwidth parameter corresponds to the currently used bin width in the histogram. The density is also shown in the spineplot view. The density for the selected data is scaled to the size of the selected subgroup by default. This can be switched off to compare the density functions of the complete sample and the selected subgroup. For an untransformed x-axis, a CD-plot can be chosen to visualize the conditional density.

Scatterplot

As a very generic plot, scatterplots can be plotted for variables of any scale. When plotting categorical variables, it is important to understand that for alpha-numerical data, like Type, arbitrarily assigned numerical class labels will be plotted! Purely numerical data will always be plotted at their values.

The aspect-ratio of a scatterplot can be fixed to be one. The axes of a scatterplot can be flipped. An axes switch will preserve current zoom, but delete any preceding zoom-steps.

Point-size and α-transparency (the default value depends on the size of the dataset) of the points can either be set by functions within the context menu, or by using by keyboard controls. ↑ and ↓ increase and decrease the point-size by 2 pixels. ← and → will increase and decrease the α-transparency of the points. These two parameters also work for binned scatterpots to change the bin parameters. The mode of a scatterplot is chosen automatically. Whenever it takes longer than 1 second to draw the points in a scatterplot, the plot is switched to binned mode. This can be overridden in the context menu. The color scheme of a scatterplot can be inverted.

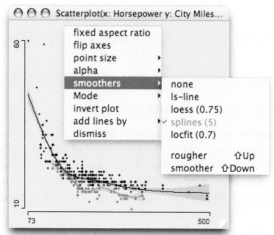

Points in a scatterplot can be joined by lines according to the grouping of a third variable. Connected lines are only shown for highlighted points.

Scatterplots can be enhanced with scatterplot-smoothers that estimate a functional relationship between the x and the y variables. There are three optional smoother functions to choose from: least-square linear fit, loess-smoother and splines. Whereas the linear fit is calculated within Mondrian, the latter two smoothers use the corresponding R functions `loess()` (with 3 robust iterations and `degree=1`) and `ns()` (from library `splines`). The smoothing parameter used is displayed in the context menu. The smoothness of the estimate can be changed by using the context menu or ↑ and ↓ in conjunction with <shift>.

Mosaic Plot and Its Variations

The mosaic plot covers the traditional mosaic plot plus three variations, namely *Same Bin-size View*, *Fluctuation Diagram* and *Multiple Barcharts*. *Double Decker Plots* can be generated by rotating all split directions to be along x. The splitting direction of the last variable can be changed (<meta>-r), and the whole plot can be rotated (<shift>-<meta>-r).

Rearranging the order of the variables is done with the four arrow keys (the order can also be directly specified by the selection order in the variable window).

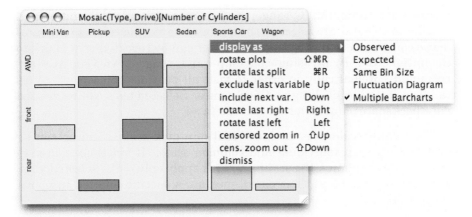

The title bar of the mosaic plot always shows the current order of the variables which are included in the plot (shown in round brackets) and the optional variables, not included in the plot (shown in square brackets). ↑ and ↓ can be used to exclude and include variables; ← and → are used to rotate the last variable in the plot with the optional variables.

Parallel Coordinate Plot & Boxplots [*y* by *x*]

Parallel Coordinate Plots (PCP) and Parallel Boxplots (PBP) have a lot in common and use the same plot-framework. It is possible to switch between PCPs, PBPs and PBPs which show the lines of the PCP only for highlighted cases.

Categorical data are tolerated in PCPs and PBPs. In the PBP-view, categorical variables are shown as a stacked barchart.

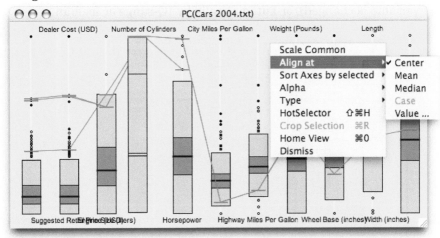

Axes in PCPs and PBPs can be reordered by <alt>-dragging an axis to its new position. Axes can also be sorted automatically. Automatic sorting includes minimum, IQ-range, median, mean, standard deviation and maximum. If cases are selected, the automatic ordering refers to the summaries of the selected subgroup. The order can be reversed or set back to the initial order. By pressing <page-up> and <page-down> the axes are permuted automatically in order to visit all possible adjacencies. Axes can be selected by clicking on their names. Selected axes can be deleted via <backspace> and inverted via <meta>-i.

The default scale uses an individual min-max scale. Scaling can be switched between individual and common scale. If single axes are selected via their names, the scaling options apply only to the selected axes. Axes can be aligned at their mean, median, a specific selected case or a certain value.

Several options are only available in PCP mode. The α-transparency of the lines can be set to six different values either via the context menu or the $\leftarrow$ and $\rightarrow$ keys. The hot-selection mode only shows the currently selected data in the PCP. The crop selection command removes the selected cases in the PCP. In any of the two modes a red frame is drawn around the plot window to warn the user that not all cases are plotted. Both modes can be switched off by the home-view function. The home-view function also fits the window to the current screen width, if an initial PCP or PBP is wider than the screen.

Boxplots y by x are quite different from the above-mentioned parallel plots, because they plot a single continuous variable split by a single categorical variable. Thus, the boxplot y by x only offers sorting options for the axes. Axes can be sorted manually via a corresponding barchart (the boxplot y by x will update automatically whenever the order of the by-variable is changed), or automatically by IQ-range or median.

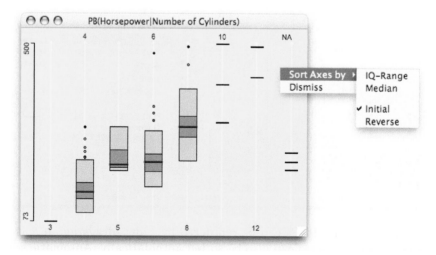

Maps

If a datafile points to additional map data, a map can be drawn for this dataset. In contrast to other plots, all controls of a map-window are gathered in a toolbar at the top of the window. Context menus are not used.

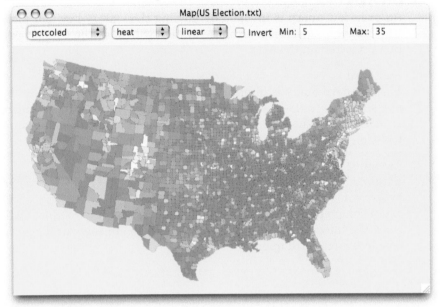

The most important control is the pop-up list to specify the variable used to color-shade the map. The next pop-up to the right gives several color scheme options, ranging from gray scales over red, green, blue and blue-to-red, to blue-white-red. The three R color-schemes `heat`, `terrain` and `topo` can also be used. All color schemes can be applied linearly, normal distributed or by rank. The color-schemes can be reversed by using the invert-option. The distribution of the variable used to color-shade the map can be limited to a specified minimum and maximum. The α-transparency of the map boundaries can be changed with $\leftarrow$ and $\rightarrow$.

Sometimes very small polygons are almost invisible. Therefore, all selected polygons can be highlighted via a red dot by pressing <meta>-f.

i.3 Conventions

Mondrian uses strict conventions to map functionality like selections, color brushing, queries or zooming to user interface controls. These controls are the same throughout the application.

Queries

One of the most important features in interactive graphics is queries. Queries in Mondrian work on three levels.

1. Overview

   An overview query is invoked by holding the <alt>- key. All plots which have a classical coordinate system will show the cursor position projected on the axes.

2. Standard

   The standard query is implemented for every plot, and triggered by pressing the <ctrl>-key and moving the mouse pointer to the object of interest. It shows the information on this object in a standard tool-tip.

3. Extended

   Extended queries are similar to standard queries. They are invoked by pressing <ctrl> and <shift> simultaneously. The result of an extended query depends on the plot type. All glyph-based plots will show the information on the variables selected in the variable window in an extended query.

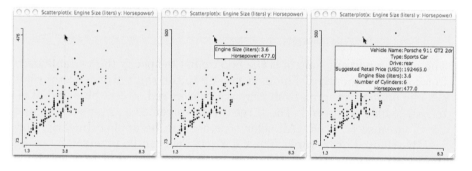

Selection

Selections can be performed in two modes: one-click selections or Selection Sequences. One-click selections can either be a single click on an object or a drag-box which encloses several objects. Selections within Selection Sequences are always represented by selection rectangles. To re-size a selections rectangle move one of the eight handles. Brushing can be performed by simply moving the selection rectangle around. All further modifications, like changing the mode or deletion, can be performed via the context menu of the selection rectangle. In Selection Sequence mode, a single click selection might be performed temporarily to explore an intermediate what-if scenario. Whenever the existing Selection Sequence is re-activated the corresponding selection is in place again.

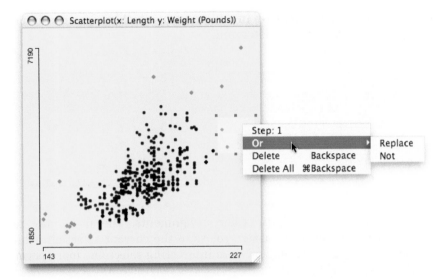

The first step of a Selection Sequence can be used to either paint or unpaint objects by choosing the OR or NOT mode of this first selection rectangle.

One-click selections only know about the current selection state, which might be combined via XOR (hold <shift>) or AND or OR (hold <shift> and <alt>; whether OR or AND is performed is set in the option menu) with a new selection. <meta>-m can be used to toggle between one-click selections and Selection Sequences.

Color Brushing

Color brushing is a persistent assignment of colors in contrast to the transient highlighting in selections. Color brushing comes in two flavors. Pressing <meta>-b in a barchart or mosaic plot will assign different colors for each category in the plot, in order to make these groups distinguishable in all other plots.

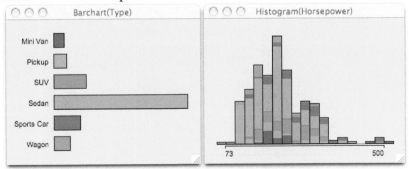

This qualitative coloring is different from the quantitative coloring generated when pressing <meta>-b in a histogram, which assigns continuous colors for each class in the histogram.

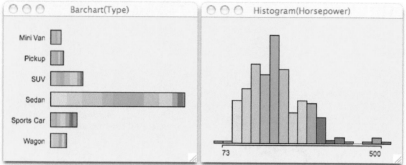

Apart from these automatic color assignments, the user may use <meta>-1 to <meta>-9 to assign colors to the current selection, which can be used to mark outliers or store interesting selections more permanently.

A color assignment can be reset by pressing <meta>-<alt>-b. Color and selection information may be derived into a categorical variable via the menu Options > Derive Variable from > ....

Zoom / Scale

Zoom operations can be performed in two ways: either by specifying a zoom region by a drag-box (use the middle mouse button under Windows and the <command>-key as modifier under MaxOS X) or by explicitly entering the new coordinate system in a dialog box. The coordinate dialog is invoked by pressing <meta>-j.

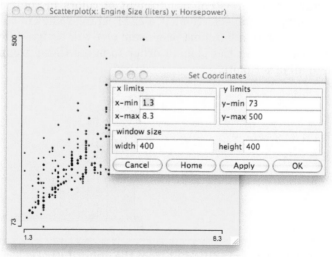

Zoom operations in Mondrian work hierarchically, i.e., a zoom-back or zoom-out operation will take the user back to the last zoom step. A zoom-out operation is performed by a zoom operation with an empty zoom-area, i.e., a middle-click under Windows and the <command>-click under MaxOS X. Depending on the plot-type, zooming is not available at all (e.g., barchart), constrained (e.g., histograms) or fully implemented (e.g., scatterplots).

Other Features

Missing Values

Missing values are coded as "NA"s in Mondrian. Missing values are not explicitly visualized as in, e.g., MANET, but they are tolerated in all views. In barcharts and mosaic plots, missing values simply form an extra group (painted in white) and integrate naturally. In histograms, scatterplots and boxplots, missing values are excluded from the data and not shown. In parallel coordinate plots missing values break the polyline. When both neighboring points are missing in a PCP, the point is still plotted as a tick on the axis. Missing values in choropleth maps result in a white coloring.

Exporting Data and Graphics

The menu-item File > Save Selection allows us to save an arbitrary subset of the current dataset. This function saves all selected cases and all variables that are selected in the variable list. An optional link to a map-file is preserved.

There are two ways of exporting graphics. The simplest way to export graphics is by using <meta>-c to copy the bitmap of the screen window to the clipboard. The graphic is copied without any modification and can be pasted into any other document that can handle images.

For a higher quality that is suitable for print, the print dialog needs to be invoked by pressing <meta>-p. After pressing <meta>-p, first the page-setup dialog is presented to specify page size and layout. In the following print dialog, the user must save the print-job to a file. Under MacOS X, this is relatively easy as the user has the option to directly save as a pdf-file or to preview the pdf and do possibly necessary crop operations. Windows users either need to choose a Postscript printer to save to a postscript file, or install an extra print-utility that allows the generation of pdf-files.

Multivariate Analysis Functions

 In Version 1.x, Mondrian supports multidimensional scaling (MDS) and principal component analysis (PCA) via the connection to R. The two-dimensional MDS is calculated via the function `sammon(d, ...)` from the `MASS`-Library. The initial configuration is calculated via a classical PCA based MDS with the command `cmdscale(d, 2)` which allows us to include missing values. The distance matrix `d` is obtained by `dist(scale(...))`. Zero distances are set to `NA`.

Whenever there are 3 or more continuous variables selected, an MDS can be calculated. The result is displayed in a scatterplot of the two dimensions of the configuration. The resulting variables are added to the variable list.

Principal components can be calculated for any set of two or more continuous variables. When $k \geq 2$ variables $x1, \ldots xk$ are selected, the corresponding k principal components `pc 1, ..., pc k` are calculated and added to the variable window. In order to support the handling of missing values, the R S3 method `princomp` for class 'formula' is used. The resulting principal components are calculated via `predict(princomp(...))` with the option `na.action = na.exclude` which will handle missing values properly.

A confirmation dialog will prompt the user whether or not the principal components shall be calculated on standardized data, i.e., a decomposition of the correlation matrix, or not. Simply hitting enter will default to the use of the correlation matrix.

Menus

Creating a plot or calculation in Mondrian always needs to specify the variables which should be included in the plot or calculation first. This follows the *object-verb-paradigm*. Depending on the currently selected variables some plots/options may be grayed out. For instance, when an alpha-numeric variable is selected, only a barchart can be plotted for the data (if the dataset includes missing values, a missing value plot can also be created). Vice versa, it does not make sense to plot a barchart for a continuous variable.

Mondrian features the following menus (also shown in Figure i.2):

- **File**

    - Open → opens a datafile in tab-separated text-file with variable headers
    - Open Database → opens a database connection (professional series only)

     – Save → saves the current datafile in tab-separated text-file with variable headers

     – Save Selection → saves the currently selected variables and cases in a datafile

     – Close Dataset → closes the dataset and all corresponding plot windows

• Plot

     – Missing Value Plot → creates a missing value plot of all selected variables which include missing values

     – Barchart → creates barcharts for all selected categorical variables

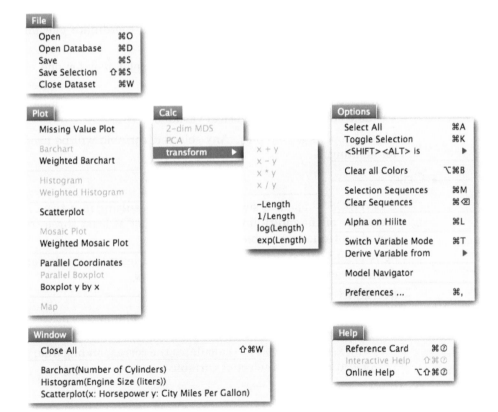

FIGURE i.2

All menus in Mondrian. Note that depending on the selection states in the variable windows some menu items are deactivated.

- Weighted Barchart → creates weighted barcharts for all selected categorical variables weighted by the exactly one selected continuous variable
- Histogram → creates histograms for all selected continuous variables
- Weighted Histogram → creates weighted histograms for all selected continuous variables weighted by a continuous variable that will be determined by a dialog box; if only one continuous variable is selected, it is weighted by itself
- Scatterplot → opens a scatterplot of any two selected variables
- Mosaic Plot → creates a mosaic plot for the selected categorical variables
- Weighted Mosaic Plot → creates a weighted mosaic plot for the selected categorical variables weighted by exactly one selected continuous variable
- Parallel Coordinates → creates a parallel coordinate plot for all (at least two) selected variables
- Parallel Boxplot → creates a parallel boxplot plot for one or more selected variables
- Boxplot y by x → creates a boxplot y by x for one continuous variable split by one categorical variable
- Map → opens a map plot. Any selected variable will be used for shading

• Calc

- 2-dim MDS → opens a scatterplot of the resulting first two dimensions of the multidimensional scaling of at least three selected continuous variables.
- PCA → adds the principal components of at least three selected continuous variables to the variable window
- **transform**
    * +, -, *, / → calculates the corresponding binary operation of the two selected variables
    * -x, 1/x, log(x), exp(x) → calculates the corresponding unary operation for the selected variable

• Options

- Select All → selects all cases of a dataset
- Toggle Selection → inverts the selection state of all cases
- **<SHIFT><ALT> is**

> * OR Selection → sets <SHIFT><ALT>-selection to OR-mode
>
> * AND Selection → sets <SHIFT><ALT>-selection to AND-mode

- Clear all Colors → removes all color assignments

- Selection Sequences → toggles between Selection Sequences and One-Click-Selection

- Clear Sequences → removes all selections from a Selection Sequence

- Alpha on Hilite → switches α-blending for highlighting on / off

- Switch Variable Mode → switches the selected non-alpha-numerical variables between categorical and continuous interpretation

- **Derive Variable from**

  * Selection → creates a binary variable reflecting the current selection

  * Color → creates a multinomial variable reflecting the current color assignment

  * Missings → creates binary variables reflecting missingness for all currently selected variables

- Model Navigator → spins off a Model Navigator to set up log-linear models using mosaic plots

- Preferences ... → opens the preference pane to set the color schemes

• Window

- Close All → closes all windows of a dataset except the variable window

- ... → any window will be registered in the window menu and show the active selection steps if Selection Sequences are used

• Help

- Reference Card → opens a window showing the reference card with all keyboard shortcuts

- Interactive Help → if switched on, a tooltip will show all options available in a plot window

- Online Help → Opens the Internet site `http://www.rosuda.org/Mondrian`

The menus in Mondrian are stateless and offer only functionality that is relevant for all plots. All plot specific options are accessed through context menus within the different plot windows. Most functions in the menus and context menus have keyboard shortcuts that are listed in the menus as well as in the reference card (see Section i.4).

i.4 Reference Card

The reference card sums up all keyboard and mouse shortcuts that are either universal for all plots, or special functions in specific plots. The symbols translate as follows:

⌘ — the meta-key, i.e., <ctrl> under Windows

⌥ — the <alt>-key

⇧ — the <shift>-key

⌫ — the <backspace>-key

⇞ / ⇟ — the <page-up>-key and the <page-down>-key

✳ — a single mouse-click

If ⌘, ⌥ or ⇧ are used in conjunction with a letter-key or the mouse, they are meant as a modifier that is pressed simultaneously with the letter-key or the mouse-click.

All Plots

⌘ A Select all cases (regardles of current selection)
⌘ B Color brush cases (all rectangle based plots)
⌥ ⌘ B Clear all assigned colors
⌘ C Copy graphics to clipboard (needs Java 1.4.+)
⌘ J Modify coordinate system of current plot (if applicable)
⌘ K Toggle to complement of current selection
⌘ L Toggle α-transparency for highlighting on/off
⌘ M Toggle between simple selection and Selection Sequences
⌘ P Print (save as '.pdf' or '.ps' in print dialog box)
⌫ Delete active selection in window
⌘ ⌫ Delete complete selection sequence
⌘ 0 Home View (Histograms and PCPs only)
⌘ 1…9 Assign predefined color 1 to 9 to currently selected cases
\<ctrl> Query object on mouse-over (add ⇧ for extended query)
⌥ Show coordinates of cursor (if applicable)
✳ Create selection (drag for selection rectangle)
  hold ⇧ for XOR-, ⇧ and ⌥ for AND/OR-selection
⌘ ✳ Zoom in (drag box)/out (single click) (middle ✳ on Windows)

Barchart

⌘ R Toggle between Barchart and Spineplot
⇟/⇞ Select next/previous category
⌥✳ Drag bar between or onto new position(s)

Mosaic Plot

⌘ R Rotate last variable
⇧⌘ R Rotate complete plot
↓/↑ Include/Exclude next variable
→/← Exchange with last/first optional variable
⇧↓/⇧↑ Decrease/Increase cell sizes (censored zooming)

Histogram

⌘ R Toggle between Histogram and Spinogram
⌘ D Switch density estimation on/off (needs R)
⌘ E Toggle between density and conditional prob. (needs R)
↓/↑ Increase/Decrease number of bins
→/← Move anchorpoint right/left

Map

⌘ F Highlight selected polygons with red circle
→/← Increase/Decrease α-transparency of boundaries

Scatterplot

↓/↑ Decrease/Increase size of points/bins
→/← Increase/Decrease α-transparency
⇧↓/⇧↑ Decrease/Increase roughness of smoother (needs R)

Parallel Coordinate Plot

⌘ R Crop selected lines (PCP view only)
⇧⌘ H Switch hot selection mode on/off
⌘ I Invert selected axes
⌫ Delete selected axes
→/← Increase/Decrease α-transparency
↓/↑ Decrease/Increase scaling of axes (aligned axes only)
⇟/⇞ Cycle through axes permutations
⌥✳ drag axes to new position

References

Azar, O. H. (2002). The social norm of tipping: A review, *Others 0309006*, EconWPA. available at http://ideas.repec.org/p/wpa/wuwpot/0309006.html.

Becker, R. A. and Cleveland, W. S. (1987). Brushing scatterplots, *Technometrics* **29**(2): 127–142.

Becker, R. A., Cleveland, W. S. and Shyu, M. (1996). The visual design and control of trellis displays, *Journal of Computational and Graphical Statistics* **6**(1): 123–155.

Bertin, J. (1983). *Semiology of Graphics*, 2nd edn, University of Wisconsin Press, Madison.

Bryant, P. G. and Smith, M. A. (1995). *Practical Data Analysis: Case Studies in Business Statistics*, Richard D. Irwin Publishing, Monewood, IL.

Carr, D. B., Littlefield, R. J., Nicholson, W. L. and Littlefield, J. S. (1987). Scatterplot matrix techniques for large n, *Journal of American Statistics Association* **82**(398): 424–436.

Cleveland, W. S. (1979). Robust locally weighted regression and smoothing scatterplots, *Journal of the American Statistical Association* **74**(368): 829–836.

Cleveland, W. S. (1985). *The Elements of Graphing Data*, Wadsworth, Monetrey, CA.

Cleveland, W. S. (1988). Graphics: 1965–1985, *The collected works of John W. Tukey*, Chapman and Hall, New York.

Cleveland, W. S. (1993). *Visualizing Data*, Hobart, Summit, NJ.

Cleveland, W. S. and McGill, M. E. (1988). *Dynamic Graphics for Statistics*, Wadsworth and Brooks/Cole, Pacific Grove, CA.

Conlin, M., Lynn, M. and O'Donoghue, T. (2003). The norm of restaurant tipping, *Journal of Economic Behavior & Organization* **52**(3): 297–321. available at http://ideas.repec.org/a/eee/jeborg/v52y2003i3p297-321.html.

Cook, D., Caragea, D. and Honavar, V. (2004). Visualization in classification problems, *Proceedings of the 2004 CompStat*, Springer, Heidelberg, pp. 823–830.

Cook, D. and Swayne, D. F. (2007). *Interactive and Dynamic Graphics for Data Analysis: With Examples Using R and GGobi*, Springer, New York.

Cox, D. R. and Snell, E. J. (1981). *Applied Statistics — Principles and Examples*, Chapman and Hall, London.

Fienberg, S. (1985). *The Analysis of Cross-Classified Categorical Data*, The MIT Press, Cambridge & London.

Forina, M., Armanino, C., Lanteri, S. and Tiscornia, E. (1983). Classification of olive oils from their fatty acid composition, *in* H. Martens and H. Russwurm (eds), *Food Research and Data Analysis*, Applied Science Publishers, London UK, pp. 189–214.

Friendly, M. (1994). Mosaic displays for multi-way contingency tables, *Journal of the American Statistical Association* **89**: 190–200.

Hastie, T., Tibshirani, R. and Friedman, J. (2001). *The Elements of Statistical Learning*, Springer, New York.

Heiberger, R. M. and Holland, B. (2004). *Statistical Analysis and Data Display: An Intermediate Course with Examples in S-PLUS, R, and SAS*, Springer, New York.

Hofmann, H. (2001). *Graphical Tools for the Exploration of Multivariate Categorical Data*, Books on Demand, Norderstedt.

Hofmann, H. and Theus, M. (under revision). Interactive graphics for visualizing conditional densities, *Journal of Computational and Graphical Statistics* .

Hout, M., Mangels, L., Carlson, J. and Best, R. (2004). Working paper: The effect of electronic voting machines on change in support for Bush in the 2004 Florida elections, *Unpublished paper* .

Inselberg, A. (1985). The Plane with Parallel Coordinates, *The Visual Computer* **1**: 69–91.

Kosslyn, S. M. (1994). *Elements of GRAPH DESIGN*, Freeman and Company, New York.

Meyer, D., Zeileis, A. and Hornik, K. (2007). Visualizing multi-way contingency tables, *in* C.-h. Chen, W. Härdle and A. R. Unwin (eds), *Handbook of Data Visualization*, Springer Handbooks of Computational Statistics, Springer, Heidelberg.

Nolan, D. and Speed, T. (2000). *Stat Labs: Mathematical Statistics Through Applications*, Springer, New York.

Oken, E., Kleinman, K. P., Rich-Edwards, J. and Gillman, M. W. (2003). A nearly continuous measure of birth weight for gestational age using a United States national reference, *BMC Pediatrics* **3**(6).

Pukelsheim, F. (2006). *Optimal Design of Experiments*, Vol. 50 of *Classics in Applied Mathematics*, Society for Industrial and Applied Mathematics, Philadelphia, PA.

R Development Core Team (2006). *R: A Language and Environment for Statistical Computing*, R Foundation for Statistical Computing, Vienna, Austria. ISBN 3-900051-07-0.
URL: *http://www.R-project.org*

Ries, P. N. and Smith, H. (1963). The use of chi-square for preference testing in multidimensional problems., *Chemical Engineering Progress* **59**: 39–43.

Robbins, N. A. (2005). *Creating More Effective Graphs*, Wiley, New York.

Scott, D. (1992). *Multivariate Density Estimation — Theory, Practice, and Visualization*, Wiley, New York, NY.

Shneiderman, B. (1996). The eyes have it: A task by data type taxonomy for information visualizations, *VL '96: Proceedings of the 1996 IEEE Symposium on Visual Languages*, IEEE Computer Society, p. 336.

Swayne, D. F., Cook, D. and Buja, A. (1998). Xgobi: Interactive dynamic data visualization in the x window system, *Journal of Computational and Graphical Statistics* **7**(1): 113–129.

Swayne, D. F., Lang, D. T., Buja, A. and Cook, D. (2003). Ggobi: Evolving from xgobi into an extensible framework for interactive data visualization, *Computational Statistics & Data Analysis* **43**(4): 423–444.

Theus, M. (1996). *Theorie und Anwendung Interaktiver Statistischer Graphik*, number 14 in *Augsburger mathematisch-naturwissenschaftliche Schriften*, Wissner, Augsburg.

Theus, M. (1999). Trellis displays, *in* S. Kotz and C. Read (eds), *Encyclopedia of Statistical Science, Update Volume III*, Wiley, New York.

Theus, M. (2003). Interactive data visualization using Mondrian, *Journal of Statistical Software* **7**(11): 1–9.

Theus, M. (2007). High dimensional data visualization, *in* C.-h. Chen, W. Härdle and A. R. Unwin (eds), *Handbook of Data Visualization*, Springer Handbooks of Computational Statistics, Springer, Heidelberg.

Theus, M. and Lauer, S. (1999). Visualizing loglinear models, *Journal of Computational and Graphical Statistics* **8**(3): 396–412.

Theus, M. and Urbanek, S. (2004). iplots: Interactive Graphics for R, *Statistical Computing and Graphics Newsletter* **15**(1).

Tierney, L. (1991). *LispStat: An Object-Orientated Environment for Statistical Computing and Dynamic Graphics*, Wiley, New York.

Tufte, E. R. (1983). *The Visual Display of Quantitative Information*, Graphic Press, Cheshire, CT.

Tufte, E. R. (1997). *Visual Explanations*, Graphic Press, Cheshire, CT.

Tukey, J. W. and Tukey, P. (1990). Strips displaying empirical distributions: Textured dot strips, *Technical report*, Bellcore Technical Memorandum.

Tukey, J. W. and Wilk, M. B. (1965). *Data Analysis and Statistics: Techniques and Approaches*, California Institute of Technology, Pasadena, CA.

Unwin, A. R., Hawkins, G., Hofmann, H. and Siegl, B. (1996). Interactive graphics for data sets with missing values - manet, *Journal of Computational and Graphical Statistics* **5**(2): 113–122.

Unwin, A. R. and Hofmann, H. (1988). New interactive graphics tools for exploratory analysis of spatial data, *in* S. Carver (ed.), *Innovations in GIS No.5*, Routledge, Taylor & Francis Group, London.

Unwin, A. R., Theus, M. and Hofmann, H. (2006). *Graphics of Large Datasets — Visualizing a Million*, Springer, New York.

Urbanek, S. (2006). *Exploratory Model Analysis — An Interactive Framework for Model Comparison and Selection*, Books on Demand, Norderstedt.

Urbanek, S. and Theus, M. (2003). iPlots — High Interaction Graphics for R, *Proceedings of the DSC 2003 Conference*.

Velleman, P. F. (1997). *DataDesk Version 6.0 — Statistics Guide*, Data Description Inc., Ithaca, NY.

Venables, W. N. and Ripley, B. D. (1999). *Modern Applied Statistics with S+*, 3rd edn, Springer, New York.

Wainer, H. (2004). *Graphic Discovery: A Trout in the Milk and Other Visual Adventures*, Princeton University Press, Princeton, NJ.

Wallgren, A., Wallgren, B., Persson, R., Jorner, U. and Haaland, J.-A. (1996). *Graphing Statistics and Data*, Sage Publications, Newburg Park, CA.

Wegman, E. J. (1990). Hyperdimensional data analysis using parallel coordinates, *Journal of American Statistics Association* **85**: 664–675.

Wilkinson, L. (1999). Dotplots, *The American Statistician* **53**(3): 276–281.

Wilkinson, L. (2005). *The Grammar of Graphics*, 2 edn, Springer, New York.

Wills, G. (1996). Selection: 524,288 ways to say "this is interesting", *Proceedings of IEEE InfoVis '96*, pp. 54–60.

Young, F. W., Valero-Mora, P. and Friendly, M. (2006). *Visual Statistics: Seeing Data with Dynamic Interactive Graphics*, Wiley, New York.

Zupan, J. and Gasteiger, J. (1999). *Neural Networks in Chemistry and Drug Design*, John Wiley & Sons, Inc., New York.

Author Index

Index

T - #0122 - 111024 - C292 - 234/156/14 - PB - 9780367452537 - Gloss Lamination